Beyond Biopolitics

"Confronting the violence that 'pulverizes being' Debrix and Barder invite us to expand our attachment to biopolitics and to take in the 'horror' of contemporary violence, understanding that what is at stake is not the end of human life itself but the end of the human condition. Mobilizing a host of philosophers from Butler, Agamben and Mbembe to Esposito and Cavarero, they urge us to reckon with power as something that kills. This is a valuable, provocative and passionate intervention."

Sherene H. Razack, *University of Toronto, Canada.*

"Why must we move beyond biopolitics? Debrix and Barder assail the foundations of biopolitical critique, showing how such approaches remain wedded to a liberal model of life-affirmation which blinds them to central aspects of death and destruction. Pushing their analysis beyond the limits of human life, Debrix and Barder show how 'agonal sovereignty' mobilizes fear, flesh, and failure to erase those modes of being whose singularity resists the logics of contemporary international politics. Drawing on such diverse sites of terror as Mexican narco killings, the Guantanamo concentration camp, and the transmission of transnational disease, *Beyond Biopolitics* identifies the complex economies of horror underpinning international relations, while also providing an accessible and critical reading of Arendt, Schmitt, Mbembe, Agamben, and Foucault."

Kennan Ferguson, *University of Wisconsin–Milwaukee, USA.*

Beyond Biopolitics exposes the conceptual limits of critical biopolitical approaches to violence, war, and terror in the post-9/11–War on Terror era.

This volume shows that such popular international political theories rely upon frames of representation that leave out of focus a series of extreme forms of gruesome violence, which have no concern for the preservation of life, a crucial biopolitical theme. Debrix and Barder mobilize different concepts—horror, agonal sovereignty, the pulverization of the flesh, or the notion of an inhumanity-to-come—to shed light on past and present ghastly scenes and events of violence that seek to undo the very idea of humanity. To highlight the capacity of horror to be in excess of both violence and the meaning of humanity, *Beyond Biopolitics* provides a series of engagements with issues much debated in contemporary critical theory circles, in particular war and terror, the production of fear, states and spaces of exception, and alterity as enmity.

This work will be of great interest to scholars of critical international relations theory, critical security studies, and international relations.

François Debrix is ASPECT Director and Professor of Political Science at Virginia Tech., USA.

Alexander D. Barder is a Doctoral Candidate in the Department of Political Science at Johns Hopkins University, USA.

Interventions

Edited by: Jenny Edkins
Aberystwyth University
and Nick Vaughan-Williams,
University of Warwick

'As Michel Foucault has famously stated: "knowledge is not made for understanding; it is made for cutting." In this spirit The Edkins–Vaughan-Williams Interventions series solicits cutting edge, critical works that challenge mainstream understandings in international relations. It is the best place to contribute post disciplinary works that think rather than merely recognize and affirm the world recycled in IR's traditional geopolitical imaginary.'

Michael J. Shapiro, University of Hawai'i at Mãnoa, US

The series aims to advance understanding of the key areas in which scholars working within broad critical post-structural and post-colonial traditions have chosen to make their interventions, and to present innovative analyses of important topics.

Titles in the series engage with critical thinkers in philosophy, sociology, politics and other disciplines and provide situated historical, empirical and textual studies in international politics.

Critical Theorists and International Relations
Edited by Jenny Edkins and Nick Vaughan-Williams

Ethics as Foreign Policy
Britain, the EU and the other
Dan Bulley

Universality, Ethics and International Relations
A grammatical reading
Véronique Pin-Fat

The Time of the City
Politics, philosophy, and genre
Michael J. Shapiro

Governing Sustainable Development
Partnership, protest and power at the world summit
Carl Death

Insuring Security
Biopolitics, security and risk
Luis Lobo-Guerrero

Foucault and International Relations
New critical engagements
Edited by Nicholas J. Kiersey and Doug Stokes

International Relations and Non-Western Thought
Imperialism, colonialism and investigations of global modernity
Edited by Robbie Shilliam

Autobiographical International Relations
I, IR
Edited by Naeem Inayatullah

War and Rape
Law, memory and justice
Nicola Henry

Madness in International Relations
Psychology, security and the global governance of mental health
Alison Howell

Spatiality, Sovereignty and Carl Schmitt
Geographies of the Nomos
Edited by Stephen Legg

Politics of Urbanism
Seeing like a city
Warren Magnusson

Beyond Biopolitics
Theory, violence and horror in world politics
François Debrix and Alexander D. Barder

The Politics of Speed
Capitalism, the state and war in an accelerating world
Simon Glezos

Beyond Biopolitics

Theory, violence, and horror in world politics

François Debrix and Alexander D. Barder

Routledge
Taylor & Francis Group
LONDON AND NEW YORK

First published 2012
by Routledge
2 Park Square, Milton Park, Abingdon, Oxon, OX14 4RN

Simultaneously published in the USA and Canada
by Routledge
711 Third Avenue, New York, NY 10017

Routledge is an imprint of the Taylor & Francis Group, an informa business

First issued in paperback 2012

British Library Cataloguing in Publication Data
A catalogue record for this book is available from the British Library

Library of Congress Cataloging in Publication Data
Debrix, François.
Beyond biopolitics: theory, violence, and horror in world politics / François Debrix and Alexander D. Barder.
p. cm. – (Interventions)
Includes bibliographical references and index.
1. Biopolitics. 2. Terrorism. 3. War on Terrorism, 2001–9. 4. Political violence. I. Barder, Alexander D. II. Title.
JA80.D43 2011
327.101 – dc22 2011008643

ISBN: 978-0-415-64366-5 (pbk)
ISBN: 978-0-415-78059-9 (hbk)
ISBN: 978-0-203-80481-0 (ebk)

Typeset in Times New Roman
by Glyph International Ltd.

Contents

Acknowledgements

Since we devised this project some three years or so ago now, we have had the opportunity to present our thoughts and ideas to many friends and colleagues who have been most gracious in listening to us, spending time with us exchanging views about some of the book's key themes, and often diligently and critically reading many of the chapters. We owe an enormous debt of gratitude to Clair Apodaca, Mat Coleman, Simon Dalby, Clem Fatovic, Kennan Ferguson, Harry Gould, Jairus Victor Grove, Nick Kiersey, Mark Lacy, Tim Luke, Scott Nelson, Nick Onuf, Brent Steele, Cindy Weber, and Jason Weidner for providing us encouragements and crucial insights along the way. At various professional conferences and invited lectures and presentations, we gathered important critical feedback from Rick Ashley, Jane Bennett, Didier Bigo, Peter Breiner, David Chandler, Jenny Edkins, Kathy Ferguson, David Grondin, John Heathershaw, Bettina Koch, Stephen Legg, Claudio Minca, Francis Moreault, Louiza Odysseos, David Pan, Mustapha Pasha, Rory Rowan, Marcy Schnitzer, Giorgio Shani, Michael Shapiro, Nevi Soguk, Jean-François Thibault, Gerard Toal, Hiroyuki Tosa, Andrew Valls, Nick Vaughan-Williams, Luc Vigneault, Rob Walker, Edward Weisband, and Laura Zanotti. They kept us intellectually honest, and they motivated us to push our critical arguments further. A special thanks goes to all the participants of the international relations working group at Johns Hopkins University, including Kellan Anfinson, Willy Blomme, Derek Denman, Stefanie Fishel, Nicole Hughes, Hitomi Koyama, Renee Marlin Bennett, Casey McNeil, and Ben Meiches. They were always eager to hear our thoughts and to engage our perspectives every time we were lucky to present our work to them. Three anonymous reviewers for Routledge offered us useful comments and suggestions that strengthened the manuscript. We are grateful for their reviews. We would also like to thank Jenny Edkins and Nick Vaughan-Williams, the editors of the "Interventions" series, who supported the project from the very beginning and were always available for sage advice and necessary analytical input. Finally, we wish to recognize Craig Fowlie and Nicola Parkin for their unwavering support, their patience, and their strong belief in the importance of critical thinking.

Initial thoughts further developed in Chapter 1 were published in French in the journal *Études Internationales*, Vol. 40, No. 1 (2009) (also set to be reproduced in English in a forthcoming issue of *Philosophy and Social Criticism*). An earlier version of Chapter 2 appeared in *International Political Sociology*, Vol. 3, No. 4 (2009). And a substantially modified and shorter version of Chapter 3 appeared in Stephen Legg (ed.), *Spatiality, Sovereignty and Carl Schmitt: Geographies of the Nomos* (London: Routledge, 2011). Our thanks go to the editors of these publications and to Stephen Legg for allowing us to share our work with their readers.

Alex would like to dedicate the book to his wife, Alla Mirzoyan. François dedicates the book to his wife and beloved colleague, Clair Apodaca, and to his grandson (and no doubt future colleague) Jeremy White.

Introduction

Beyond biopolitics

> For a long time some people believed that if the horror could be made vivid enough, most people would finally take in the outrageousness, the insanity of war.
>
> Susan Sontag[1]

> Only modernity makes of individual self-preservation the presupposition of all other political categories, from sovereignty to liberty.
>
> Roberto Esposito[2]

> Humans have long been distinguished from other animals by our ability to create. Our distant ancestors learned how to tame the wild, reach the top of the food chain, and build civilization. Our more recent forbears figured out how to crack the codes of science, and even escape the bonds of gravity, taking our species beyond our home planet … And now we are creating something exciting and new, a technology that might just transform humans' role in their world, perhaps even create a new species.
>
> P. W. Singer, *Wired for War*[3]

The biopolitical framing of life and death

Since the United States launched its Global War on Terror in the immediate aftermath of the 9/11 attacks, visual representations of war, its atrocities, and its agony have been defining moments in critical political reflections over the relationship between terror and violence. Whether we think about the live images of the Twin Towers' collapse on 9/11, the delayed visual feeds of US troops "shocking and awing" their way into Baghdad in March 2003, the "mission accomplished" televised speech by then US President George W. Bush on the deck of the USS Abraham Lincoln a few weeks later, or the snapshots of tortured and humiliated Iraqi prisoners in Abu Ghraib in 2004, the visual image has been a close partner to—and perhaps a participant in—our contemporary understanding of war and its supposed objectives. It is in the context of such a multiplication of instances of representation of war by way of visual media that, in a recent essay, philosopher Judith Butler offers to engage the late photography theorist Susan Sontag in an inquiry over the meaning of war violence, torture, and the photographic image.[4] Exploring how what she calls "frames of representability" are linked to broader social and political norms about what of late has been made to count as "grievable life" and what has not, Butler finds herself confronted with Sontag's famous claim that, in and of itself, the photographic image does not offer any sort of interpretation or understanding. According to Sontag, photography is crucial in providing affect, a sentimental response, or necessary emotional reactions to the atrocious events it witnesses and captures. But, short of any accompanying narrative, a photograph remains fragmentary in relation to meaning and understanding.[5] The bits and pieces of "real life" situations photography provides the viewer with lack the narrative coherence and constancy Sontag believes is necessary for proper representation and realization.

Butler counters Sontag's perspective by claiming that, on the contrary, the photographic frame itself already contains and enables certain interpretations. The photographic image, Butler suggests, is as much about what is contained in the image as it is about what is outside the frame of the photo, and thus kept unnoticed, unseen, invisible, or occluded (purposefully or not). As such, she goes on to state that "interpretation is not to be conceived restrictively in terms of a subjective act … [since it] takes place by virtue of the structuring constraints of genre and form on the communicability of affect."[6] Butler further argues that "it is not just that the photographer and/or the viewer actively and deliberately interpret, but that the photograph itself becomes a structuring scene of interpretation."[7] While photographic images may lack narrative coherence, they still have a way of framing and shaping reality, often through the role and power of affect and sentiments. Framing, or better yet what fits in the visual frame (as well as what is not included in it), is already an act of interpretation and understanding. The photographic image thus already comes to us with certain embedded rules and principles of intelligibility, or what Butler refers to as a "field of representability." Any field of representability, Butler believes, is constituted as much by what is left out as by what the photo appears to show or reveal. This is what leads Butler to provocatively claim that "the photograph interprets us."[8]

Sontag would probably object to such a statement. For Sontag, the photographic image never interprets us. At best, it affects us. The role of affect in photography is in fact far more important for Sontag than any capacity of interpretation it may or may not have. It is particularly so, Sontag goes on to affirm, for war photos. Butler indicates that, for Sontag, in times of war, the "transitive affectivity of the photograph may overwhelm and numb the viewers."[9] Numbed viewers are less likely to understand what the reality that the photo allegedly depicts is about. Their capacity of interpretation is thus further dulled, unless narrative/textual explanations or captions accompany the image. Still, for Sontag, this incapacity of war photography to achieve or enable greater meaning is not only expected but also acceptable, because war photos serve a different and crucial function. War photos, Sontag believes, are necessary to represent human suffering. They represent human suffering not in the sense that they allow the viewer to understand its origins, contexts, and connections, but rather because they establish "through the visual frame a proximity that keeps us alert to the human cost of war, famine, and destruction in places that may be distant."[10] Thus, Sontag argues that war photos fulfill an ethical duty by awakening our moral indignation.

At the same time, Sontag is also aware of the danger of loading photography, particularly war images, with affective meaning. Insisting on the importance of war photos as a result of their moral function could easily allow photography to fall into the "shock and awe" mode of apprehension. According to Sontag, shocking in photography amounts to spectacularizing a scene or fragment of reality. As Butler further notes, the "shock value" of war photography "tends to aestheticize suffering for the purposes of satisfying a consumer demand."[11] Such a tendency in war photography undoes any moral value the war image might have. Interestingly, the possibility that war photography may turn into a sensationalized "consumer product" also suggests that, despite her claim to the contrary, Sontag does recognize the importance of framing and representability (both what is in and what is left out of the picture) in the visual. Thus, Sontag's position regarding the function of the visual image may not be so removed from Butler's perspective after all. Indeed, the risk with the war image for Sontag is that the affective proximity such an image conveys may fall prey to one dominant emotional perspective or sentiment. When one such dominant sentiment prevails over the content and meaning of war photos, both the notion of photographic affect and the viewer's ability to understand reality are distorted (this is once again why Sontag prefers text over image when it comes to narrative coherence).

Thus, the concern with the war image cannot just be with what it shows, but also with "how it shows what it shows."[12] Whereas Sontag only indirectly arrives at such a conclusion and implicitly agrees that a focus on the "how" of photography is just as crucial as an insistence on the "what," Butler wholeheartedly subscribes to such a view. Butler writes that "the 'how' not only organizes the image, but works to organize our perception and thinking as well."[13] For war photos then, what Butler calls the "operation of the frame" is key. How war photos show what they show, what sort of dominant sentimental or cognitive interpretative structures delimit the field of vision, and what fragments of reality are already interpreted for us (or, better yet, already interpret us) are urgent questions that, for different reasons, both Butler and Sontag ask us—viewers/consumers of a steady flow of war images (both before and after the invention of the Global War on Terror)—to pose. Through those questions that interrogate not just the contents of war images, nor even simply the framing of those photos, but rather the constant complex relation between frame and content, we are being asked to explore the production of reality, affect, and meaning at the level of a field of representability, that is to say, through conditions and structures of visibility and invisibility, of the seen and the unseen, that enable certain representations of reality to the detriment of others. As some representations are in focus while others are placed out of sight, the reality and affective value granted to certain human bodies and the ability of certain lives to be grieved (as Butler would put it) shift accordingly.

The concern with the "how" as much as the "what" of the war image lies at the heart of photojournalist Peter van Agtmael's recent collection of photos from the US wars in Iraq and Afghanistan, *2nd Tour, Hope I don't Die.*[14] From 2006 to 2008, van Agtmael was embedded with US troops in Iraq and Afghanistan, sometimes following the troops and their families back home in the United States. Van Agtmael's collection is made up of various snapshots of everyday scenes from Iraq, Afghanistan, and the United States. Some of those scenes are very mundane, non-eventful, while a few others are more dynamic, showing soldiers or local Iraqi and Afghan people in action. Still, the bulk of the collection (and the volume that reproduces these photos) is about US troops, often individual soldiers, though they rarely are posing for the camera. The soldiers are shown in their apparent "individuality," as human beings or singular lives, going about their daily tasks or routines. These tasks or routines cover anything from searching and rand-sacking an Iraqi family's house in Mosul, resting on a cot late at night and watching movies on a DVD player while on a mission in a remote eastern Afghan province, or falling asleep in one's recliner in the middle of the day at home in Wisconsin, with painkillers and cigarettes nearby, having been honorably discharged after a rocket attack blew most of one's leg, to trying to stay conscious and alive onboard a medevac after having been shot in the leg, stomach, or face during a patrol in what could have been either Iraq or Afghanistan, taking a brief break and gathering oneself in the ER of Baghdad's Combat Support Hospital after one was unable to save the life of another soldier who came to the operating bloc badly burned or disfigured, or paying one's respects in front of a recently fallen comrade's makeshift grave in Iraq made up of his boots, his picture, his helmet lying on top of his rifle, and the US flag.

In a move that Sontag probably would have appreciated, most of the photos in the volume come with captions or longer texts. Thus, one is led to believe that the images of war contained in this collection are not just meant to shock or evoke a moral reaction. They also explain, and favor interpretation and understanding. From the opening page of the volume, even before the first image is displayed, van Agtmael tells us that the photos do come with a particular frame and that they are laden with a certain interpretation. Here, interpretation and the role of the visual frame are very much subjective. The photos seem to reflect or represent the photographer's own perspective. Van Agtmael writes: "I wanted to make pictures that reflected my complex and often contradictory experiences, where the line was continuously blurred between perpetrator and victim, between hero and villain."[15] Thus, the war photos in *2nd Tour, Hope I don't Die* are as much about the individuality of the frame itself (here, of the photographer's own visual experience) as they are about the individuality of the people, mostly soldiers, being shown. Or rather, they are about the connection or transposition of the photographer's subjectivity and identity into the singular images—often bodies and faces—of the people that his camera captures. At this level at least, van Agtmael's images operate within a frame of representability that enables affect far more than understanding.

Again, it is likely that Sontag would have appreciated such a use of the war image. Van Agtmael's photos rarely shock. Or, to the extent that they do, it is not because the suffering and war atrocities they reveal are embellished or sensationalized. Again, many of the scenes depicted and framed by van Agtmael's camera are very much banal and non-spectacular. There is blood, there are wounds, there is injustice, there is pain, and there is destruction and terror in many of those photos. But all of these characteristic marks of war images are rather contained or limited here. Van Agtmael even admits that, faced with a moment of death (of a soldier or of a civilian victim) or with horrendous scenes of mutilation or dismemberment, he often chose to turn the camera away or off. When van Agtmael's photos create shock and are able to achieve "transitive affectivity," it is when they become evidentiary materials, when they reveal pain and suffering as the daily costs of war. Here, aesthetic effects are not required to arrive at this degree of affect. While such photos, Sontag would surely claim, still escape explanation and understanding (although, in van Agtmael's collection/volume, the accompanying texts do compensate for this lack), they exude authenticity and look like they can be trusted, despite the fact that the photographer framed them this way. As Sontag puts it: "[p]ictures of hellish events seem more authentic when they don't have the look that comes from being 'properly' lighted and composed." She adds: "By flying low, artistically speaking, such pictures are thought to be less manipulative—all widely distributed images of suffering now stand under that suspicion—and less likely to arouse facile compassion and identification."[16]

Resisting facile identification between the war victim or the warrior on one side and the viewer on the other does seem to be an important element in van Agtmael's so-called subjectively selected frame of vision and representability. While, as suggested above, there appears to be an intention on the part of the photographer to transfer his own subjectivity into that of the people he shows in his photos (he interprets the images), or, rather, the civilians and soldiers' faces and bodies he captures with his camera become a part of him (the image interprets him), such a form of identification is far from obvious or simple. Confusion over one's subjectivity or individuality is more likely to be the kind of affective response one might offer while looking at these war photos. At least, this seems to be the response that van Agtmael himself provided after returning from Iraq and Afghanistan and looking at his photos again. Indeed, he writes: "A lot of my ideas about the world proved to be misconceived, and I began wondering why they had become beliefs at all … For a long time I felt stuck in my reflections."[17] Van Agtmael adds: "But in uncomfortable ways, my life is better than it was. I was searching for something, and I found it in war. I lost a lot too, and I wonder where the balance lies. I sometimes think that war continues only because there is euphoria as well as grief in survival, while the true victims are marginalized or dead."[18]

Van Agtmael's war images may resist "facile compassion and identification" (to use Sontag's phrase). The moral sentiments they convey may be as "complex and contradictory" as the photographer's own experiential or interpretative frame. Moreover, these war photos may also be hostile to any clear sense of understanding. While the fragments of war reality they provide can serve an evidentiary purpose (or bear witness), narrative consistency is by and large missing. Van Agtmael's texts and captions actually do not say much. For example, they do not constitute an outright criticism of the US wars in Iraq and Afghanistan. They do not seek to glorify the warrior heroes either. But the texts and the photos also do not fault the troops for war crimes, brutalities, or tortures (even when some images appear to reveal instances of excessive military force). Nor do they blame or castigate the Iraqi and Afghan populations for resisting the US military presence. Thus, while the viewer of these images is clearly given an interpretative framework, that of the photographer himself, such a frame of representability and intelligibility is, as van Agtmael admits, "an agonizingly limited one, a shred of all that has been utterly, unfathomably lost."[19] Either from the point of view of affect or from the perspective of explanation and understanding, there is not much that can be recovered from this iconography of war, destruction, and terror.

Or so it seems. There is, we believe, a cognitive and affective perspective that nonetheless prevails in van Agtmael's photos and texts. Yet, in order to be exposed to it, we need to take to heart Butler's invitation to pay attention to "how" the war photo shows us what it shows. And, in the context of this visual frame of war, being attentive to the "how" of the photo requires that we do not limit ourselves to the interpretative frame—complex, contradictory, or inconsistent as it may be claimed to be—that the photographer/camera seems to insist on leaving us with. Indeed, we want to suggest that van Agtmael's photos both provide an affective dimension that may facilitate some form of identification and enable a structure of explanation and meaning that goes beyond the punctual evidentiary mode of the image (this image shows this, and that is it), because, consciously or not, they have recourse to what we call a *biopolitical frame of representability*. Such a biopolitical frame of representability consists in facilitating and privileging modes of representation and intelligibility that, for diverse reasons and according to various senses of utility, place the affective and cognitive focus on the production, preservation, and enhancement of human lives and bodies. Following Roberto Esposito's quotation at the opening of this chapter, biopolitical frames of representability highlight the need for human self-preservation. These visually affective and cognitive frames seek to overload all categories of meaning with the requirement to protect and secure certain lives (and, sometimes, to create new and improved—or more resilient—ways of life too) even if it is at the cost of destroying other bodies/beings whose worthiness is to be doubted. Put differently, biopolitical frames of representability such as those introduced by van Agtmael's war photos reflect the at once enabling and delimiting work of a biopower that, as Michel Foucault has famously argued, operates at the level of a species' body, a "body imbued with the mechanics of life and serving as the basis of the biological processes," and whose main task is "to *foster* life or *disallow* it to the point of death."[20]

That the body—tired, mutilated, in pain, in shock, resting, in motion, proudly posing for the camera, hiding from sight, alive, dead—is prominently displayed throughout van Agtmael's war photos is an understatement. As suggested above, virtually every picture is about a soldier's or a civilian's face, body, and life (and his/her brief life story in the accompanying caption/text). More to the point, it is life understood or expressed as a succession of bodily motions and expressions that takes center stage through van Agtmael's visual frame. Life, van Agtmael seems to intimate, whether it is worth living or not, is a measure of the body's pains and pleasures. There is no such thing as a higher meaning to life here, no greater sense of being or humanity than what is already experienced as and through the life of the body. This is perhaps why van Agtmael suggests that "humanity is lost along the different ways we represent ourselves."[21] One could argue that the life/body that is featured in van Agtmael's photos and texts is a life that sits at the heart of biopolitical designs and operations. It is life as *zoē*, not as *bios*, the "simple fact of living" as opposed to the "politically qualified life," as Giorgio Agamben (going back to Aristotle) famously puts it.[22] It is "bare life," or the life one finds at the "threshold of biological modernity" (as Foucault argues), at "the point in which the species and the individual as a simple living body become what is at stake in a society's political strategies."[23] The bodies on display in van Agtmael's war photo album are those of *homines sacri* who can be killed but not sacrificed.[24] Dispensable, expendable lives, the lives of those *homines sacri* revealed by van Agtmael's camera are at once captured and excluded by contemporary power and by the biopolitical logic of the US-driven Global War on Terror.

Butler is one of several scholars who have suggested that there is a close kinship between biopolitics and war.[25] In this context, the war image may be able to show us "how 'life' itself is being defined and regenerated … within new modes of knowledge/power," and how war operates as a crucial instrument in biopolitical projects.[26] Van Agtmael's war photos provide us with a biopolitical frame of representability precisely because they try to make us understand (and affectively react to) the bare lives of soldiers and civilians in Iraq, Afghanistan, and the United States. They seek to represent lives as bodies that are worth living to the extent that those lives play a part in the promotion of the human species (many of the lives and bodies on display in van Agtmael's volume/collection fight terror or evil itself). They bring into focus bodies/lives that can be destroyed or killed through operations of sovereign power centered on the necessity of making life live (or, at least, that try to safeguard a certain way of life or a certain global body politic). Of course, as is always the case with frames of representability, they also leave out of sight, away from the camera's lens, other lives and bodies that are said to be a threat to humanity, or to the life of the species (terrorists, insurgents, or Taliban fighters are almost never shown by van Agtmael).[27] As Mick Dillon and Julian Reid would put it, van Agtmael's war photography hopes to frame what they call "biohumanity." Biohumanity is bare life placed to the service of war. Or, as they argue, "[c]ommitted to promoting and securing the life of the biohuman means … that liberal rule must be prepared to wage war not so much for the human, but on the human."[28] Within a biopolitical frame of representability (similar to the one introduced by van Agtmael's), what is at stake is not so much to "romanticize violence and war" (something van Agtmael clearly seeks to avoid).[29] It is also not to provide a detailed exposé on the atrocities and horrors of war (something that, as Sontag indicated, generally comes equipped with an assortment of spectacular and shocking images). Rather, what is at stake in biopolitical frames of representability is the necessity to place life and the body—deprived of any political qualification and presented as simple or natural life—at the center of operations of power (starting with war) that have the life of the human species as their sole referent object and, in seeking to make the biohuman live and prosper, constantly decide which lives are worth living and which are not (or which bodies can be killed).

Theorizing biopolitics

Bluntly summarized by Foucault, biopolitics is generally taken to be concerned with the power "to 'make' live and 'let' die."[30] Biopolitical frames of representability—visual or textual, representational or discursive—seek to introduce new modalities of knowledge/power that open up two main lines of inquiry and problematization within predominant perspectives on political modernity. First, a biopolitical framework of analysis and understanding intervenes at the level of sovereign power. Within modernity, biopower is said to emerge as a new or additional form of power, a power over life and the living, that leads to a reconsideration of the traditional model of centralized (and awe-inspiring) sovereign power. Second, as suggested by Foucault's statement above, biopolitics implies a different relationship between sovereignty, violence, and death. While death does not disappear from the biopolitical framework (as we clearly saw with van Agtmael's war photos), its function, utility, and positioning vis-à-vis life are recalibrated.

Nikolas Rose has suggested that biopower or biopolitics "is more a perspective than a concept."[31] By "perspective," Rose implies that what matters about the notion of biopolitics—particularly, as coined by Foucault—is that it helps to bring into focus a series of strategies and interventions of power, authority, and control at the level of life. The life in question in operations of biopower is the life of modern man. Or, as Foucault argues, "modern man is an animal whose politics place his existence as a living being in question."[32] According to Foucault, with the emergence of biopower, a shift in mechanisms of power is visible. Once again, this shift in power is marked by an insistence on "fostering life"—on the part of the state and sovereign, to start with, but also of multiple mechanisms of governance and control disseminated throughout society (or micro-powers). Or, put differently, biopolitics reveals the presence of a power that seeks to manage life and the well-being of living subjects taken to form a population. Still, as Rose adds, biopower cannot be construed only as "a multitude of attempts to manage" the life of human subjects since its purpose is also "to turn their individual and collective lives into information and knowledge, and to intervene on them."[33] Thus, biopower takes hold through the creation of new social problems (health, hygiene, life expectancy, social productivity, and so on) and through the need to tackle them by means of new techniques of knowledge and new modes of scientific rationality (statistics, demographics, economics, and so on) that can correct, improve, reform, or recalibrate those social issues at the level of the life of the population as a whole. Whether they focus on reforming, redressing, enhancing, or maximizing the life of the human by targeting individual bodies through disciplinary practices and power (something that Foucault already highlighted in *Discipline and Punish*[34]), or whether they seek to normalize the population or society in its entirety through various mechanisms and technologies that attempt to regulate the life of the human species, regimes of biopower extend the political authority of the sovereign or, rather, government to the "vital processes of the living subjects,"[35] not to the subjects as citizens of the state or as members of the sovereign compact. The referent object of biopolitics is indeed those "vital processes" of the population, society, or species.

With this biopolitics of the population, society, and the human species comes a transformation in relations and configurations of power. Summarizing Foucault's argument, Rose writes that biopolitics is not to be understood as "the implementation by the state of a single strategy of regulation."[36] Biopolitics is not about the power or control of the one and only centralized sovereign, even if it is a so-called benign sovereign that would have surrendered its power to put to death, to the benefit of a power to make life live and thrive. Rather, it is about a proliferation of practices, expert knowledges, technologies, dispositifs, and exercises of government (what Foucault at times refers to as "governmentality") that invest, dissect, enhance, and reproduce individual and collective life across the social spectrum, and often beneath and beyond the level of the state itself. Strategies of power centered around the life or being of the sovereign, the requirement to preserve the territorial integrity of the state, the need to defend the unity and identity of the political community (the good life "inside" safe borders) are no longer priorities under regimes of power that have made the management of vital processes of life their main objective. Biopolitical wars are no longer waged on behalf of the sovereign. Rather, they are "waged on behalf of the existence of everyone,"[37] or on behalf of life itself.

Drawing on this Foucaultian insight about the redistribution of relations of power (and their objectives) under biopolitical conditions, some scholars influenced by this perspective have argued that a distinction can be made between geopolitics of security and biopolitics of security. Whereas geopolitics of security are driven by the need to protect and defend the territory of the sovereign nation-state, biopolitics of security are geared towards constantly promoting and improving the conditions of life of a population or species. As Dillon puts it, "the *biopolitical problematisation of security* in terms of species existence is very different from that of the *geopolitical problematisation of security* in terms of territorial sovereignty."[38] Still, this passage, from a geopolitics of territorial sovereignty to a biopolitics of species life, is no guarantee that the mechanisms of power and governance involved in biopolitical operations—no matter how fragmented and disseminated throughout and across social orders they may be—will be more benign or less oppressive. Far from it, in fact. Rose indicates that wars conducted in the name of the existence of everyone "are bloodier than ever, and [biopolitical] regimes inflict holocausts on their own populations."[39] Moreover, while the inside/outside demarcation of modern political power is no longer dependent upon the need to defend or protect the geographical integrity of a special space (the territory of the sovereign state, the realm of domestic politics, the space of the political community, the orderly domain "within"), this inside/outside divide is nonetheless still operational in biopolitical configurations.[40] As Dillon argues: "biopolitics … requires mechanisms to differentiate species existence in ways that will usefully inform how it must best govern in order to pursue its vocation of promoting species life."[41] Differentiations, classifications, and categorizations are still crucial to biopolitical designs. In order to be better regulated and normalized, the life of the population/species must become the constant object of demarcating operations, even if the insides and the outsides produced as a result of those categorical manipulations are no longer primarily territorially defined.

One vital (literally) mechanism of demarcation of an inside and an outside in biopolitical modernity is race. Race, Dillon clarifies, "is one of the markers which biopolitically adjudicates."[42] What race adjudicates is precisely which sort of life is "biopolitically suitable life" and which other life (or ways of life) must "die in order for some other life to live."[43] By demarcating those lives (and species) worth living from those that are superfluous or worthless (and may not even be worthy of being called life as such), techniques and technologies of biopower inevitably place death—not just life—at the heart of their operations and considerations. Dillon indicates that it is race as a differentiating principle of life that serves as the inevitable connection between biopolitics and necropolitics, or what Achille Mbembe calls the "subjugation of life to the power of death."[44] This insight may appear contradictory to what Foucault seemed to suggest was the elementary motivation behind biopower: to foster life and disallow death. Yet, as Foucault and others have also revealed, the power to preserve life at all costs is inevitably coupled with the need to determine what must die. And, as Foucault further detailed in *"Society Must Be Defended"*, race "is precisely a way of introducing a break into the domain of life that is under power's control."[45] Dillon concurs with this apparently contradictory Foucaultian insight and believes that "biopolitics, necropolitics, and race are closely correlated phenomena."[46] Catherine Mills, in her study on liberal eugenics, further explains that "[r]acism justifies and makes possible the otherwise paradoxical deployment of death within a regime of biopower, and it does so precisely through a normalizing regulation that fragments 'the field of the biological that power controls,' that is, by introducing 'caesura within the biological continuum addressed by biopower'."[47] Thus, the reformulated and readjusted relations of power that unfold as a result of the proliferation of biopolitical techniques and rationalities through modern society may not bear the bloody and deadly marks of the sovereign's or state's power. They may present themselves as positive and productive exercises of maximization of a population's or a species' life capacities and potentials. Yet, they cannot invest life and living beings and produce revitalizing and regenerative effects throughout modernity without a power to mobilize death, one that indeed condemns lives to be erased and eradicated. Biopolitics, then, is inevitably also a necropolitics or a thanatopolitics.

Mbembe believes that any understanding of biopolitics that fails to account for the way biopower actually targets not life itself but rather what he calls "the creation of death worlds" will remain woefully incomplete and inadequate.[48] Such an adjustment to the biopolitical argument, or what Dillon prefers to label a "necropolitics of dead life,"[49] seeks to push the biopolitical perspective beyond Foucault's analyses. Contrary to what Foucault seems to intimate, Mbembe suggests that biopolitics does not just discriminate between forms of life and decide which ones are worth living and which lives must die in order to allow species life to be preserved or enhanced. Rather, and quite bluntly, the object of biopower is death itself, and it has always been so across various historical and geographical contexts. Biopolitics better rephrased as necropolitics, Mbembe seems to imply, is concerned with the submission of "natural or simple life" (as Agamben put it), or *zoē*, to that power that will terminate or deny it. In this manner, Mbembe's necropower is much closer to the old sovereign power to put to death or to invent a succession of threats and dangers that demand that the lives of human subjects be surrendered (if not sacrificed) to an absolute and arbitrary power of destruction (through war, terror, colonial rule, and so on). Mbembe's approach to necropolitics thus combines biopolitical frames of understanding (about the emphasis placed on dead life) with more traditional perspectives on sovereign force and violence (a power to put to death). One might suggest that, far from taking the biopolitical approach beyond Foucault, Mbembe actually returns it to a series of analyses (about war, colonial violence, the fight against terror, and so on) that insist on conceptualizing power as something that limits, stifles, enslaves, destroys, and ultimately kills, or what Foucault may prefer to call the "repressive hypothesis."[50]

Put differently, not enough emphasis may be placed in Mbembe's work on the production of dead life. Necropolitics or thanatopolitics, proponents of a Foucault-derived biopolitical perspective suggest, is not simply about the return of an absolute power of death. It is rather an analytical approach that allows one to bring into view the fact that "making life live is … a lethal business."[51] What this statement (by Dillon) reveals is not so much that a power over life has been taken over (yet again) by a power over death, but rather that biopolitics in fact relies upon a fundamental indistinction, often at the level of *zōe* (or bare and unqualified life), between life and death, or between making live and letting die. This indistinction is the basis for what Agamben believes is the immanent passage from biopolitics to thanatopolitics. As Agamben puts it: "[i]f there is a line in every modern state marking the point at which the decision on life becomes a decision on death, and biopolitics can turn into thanatopolitics, this line no longer appears today as a stable border dividing two clearly distinct zones."[52] For Agamben, it appears that not even race—that ultimate criterion of distinction and demarcation between worthwhile life and useless existence or even non-life—is capable of providing us with a clear separation between biopolitics and thanato- or necro-politics. Rather, under what Agamben would consider to be a generalized condition of exception, when the camp becomes the dominant paradigm for bare life (once again, the life that can be killed but never sacrificed), life inevitably is death and death is life. It is indeed beyond the point to insist on separating the two stages or conditions of living (or not living) and, more crucially, it is beyond the point to try to theorize biopolitics from the perspective of techniques or mechanisms that seek to make life live by mobilizing a power that inevitably puts to death. Put differently, it is neither life nor death (nor even dead life) that is the referent object of biopolitics for Agamben, but rather the very principle of indistinction between both, or what he prefers to identify as bare life or recognize in the figure of *homo sacer*, and that lies at the heart of modern and contemporary political practices of exception.[53]

Agamben's analysis seeks to advance Foucaultian biopolitical approaches. But, departing from Foucault, it also hopes to prompt new thinking (both within and beyond biopolitics) about the relationship between the power to promote the life of the species and the power to create "death worlds." Another contemporary Italian philosopher, Roberto Esposito, is also concerned with this interplay between life and death, and he suggests that a new analytical paradigm, derived from biopolitical principles, needs to be introduced. Esposito refers to this new perspective placed within a general frame of biopolitical intelligibility as the "immunitary paradigm." Immunization, Esposito argues, "posits a peculiar knot" in the analyses of "biopolitics and modernity."[54] The immunitary paradigm, Esposito insists, allows us to realize why biopolitics "continually threatens to be reversed into thanatopolitics."[55] But more than an indication of the presence of a constant threat to life in biopolitical designs (as might have been the case with Mbembe's notion of necropower), immunization is a modality of articulation of biopolitics and thanatopolitics operating together, and appearing as one indistinct notion (as Agamben suggested). Indeed, immunizing life is at once affirming life and destroying it. It is a matter of implanting in the life of the species and in live bodies elements or seeds of their future possible annihilation so that life/the body might actually be able to fight off those future threats, often by building anti-bodies or developing new structures of life. Immunization as a general modality of anticipation of bodily harms and life dangers and as a measure of auto-protection against these risks condemns life, bodies, populations, or the species to confront fragments of death, lethality, and termination of natural life (and it does so, sometimes, in violent and brutal ways) in order to force life itself to gain resilience, vigor, and vitality in return. As Esposito clarifies: "immunization is a negative form of the protection of life. It saves, insures, and preserves the organism, either individual or collective, to which it pertains, but it does not do so directly, immediately, or frontally; on the contrary, it subjects the organism to a condition that ... negates or reduces its power to expand."[56]

The purpose of the immunitary paradigm is thus to join biopolitics and thanatopolitics in the same operation, an operation that cannot really be said to either preserve life or destroy it. It is both about life and death, but the objective is also beyond preserving life or destroying it, beyond the mere joining of the power *of* life with the power *over* life, as Esposito suggests.[57] The objective made available by such a paradigm is to arrive at, and perhaps create, another way of life, or better yet, another form of being or existing. This other form of living or being is one whereby immunized life is always already prepped for its own negation at the hands of those agents (who call themselves protectors of life) whose very task is to guarantee that the life of the species will thrive and be enhanced. And this readiness of life to be auto-negated (not killed and not sacrificed either) lends itself to becoming paradigmatic to the extent that it can be turned into a model of protection, preservation, and securitization not just of individual organisms (cells infected with viruses or bacteria to see how the body reacts), but also of collective bodies, starting with an entire population or society (as we see time and again when epidemics are identified and require large scale immunization or vaccination campaigns). Moreover, this modality of construction of a way of living/being that requires auto-negation or auto-infection takes place at the level of the species and its so-called security as well. In particular, it is operative whenever sources of terror are said to threaten the very existence of a species' life (or what it means to be human) and whenever governmental or military agents in charge of the protection of the species against terror demand that those who recognize themselves as part of this species or this humanity implement self-effacing policies and norms in ordinary life conditions that will allow them to be free from terror (even if it is at the cost of surrendering their lives to a series of anti-terror measures).

In their latest book, Dillon and Reid choose to give this new way of living/being that results from the immunitary paradigm the names "emergent life" and "being-in-formation."[58] Being-in-formation or emergent life (concepts that we discuss in more details in Chapter 2) present themselves as modalities of living/being that are both a guarantee for future life to live and a threat to life and the species itself. Thus, in its being-in-formation, emergent or immunized life mobilizes both biopolitics and thanatopolitics at once. Emergent living is an object both of life creation and life negation, and, as Esposito intimates, it provides the immunitary platform from which the species is both endangered and safeguarded.[59] Here, the notion of biopolitics no longer needs race to invoke the power of death. Rather, it is life itself, even a possibly acceptable or validated racial conception of life, that is introduced as a danger to life. Racial demarcations are beyond the point with a conception of bio- or thanato-politics in which life is indistinctly about the potential for the species to thrive and about the proliferation of death worlds. As Dillon and Reid argue, it is indeed the "becoming-dangerous" of emergent life that defines this revised biopolitical perspective.[60] The political opportunities opened up by targeting life as "becoming dangerous" are quite different from those biopolitical outcomes (disciplinization, normalization, regularization) identified and studied by Foucault and many of his followers. Here, in the wake of biopolitics revisited through the paradigm of immunization and with the advent of emergent living, we are dealing with the realization that power (of *and* over life) may not just be in the process of establishing a continuum between techniques and technologies of life creation and those that are in the business of destroying life. Rather, we are witnessing the development of techniques and technologies (and strategies that support them) for which life creation and life destruction are indistinguishable outcomes and commonly acceptable objectives. Dillon and Reid recognize this crucial paradigmatic shift in the logic of biopolitics (if we still must call it biopolitics) when they state:

> biopolitics significantly re-cognizes death as it does birth. You do not therefore get to a more acceptable biopolitics by concentrating on birth rather than death. Biopolitically speaking, you are always already implicated in both from the beginning. Politically speaking, it is not a matter of which end you start from, therefore, or which process you prioritize. The military is as interested now, for example, in life-creating and life-adaptive processes as it is in killing, because … it locates the nature of the threat in the very becoming-dangerous of the vital signs of life itself.[61]

Many social and political theorists, international relations scholars, or critical political geographers have turned eagerly to the biopolitical perspective over the past decade, in order to develop an explanatory framework that could make sense of the way techniques, strategies, and apparatuses of power allegedly designed to promote life have made "the murder of the enemy [their] primary and absolute objective," often in the name "of war, of resistance, or of the fight against terror."[62] In their respective subfields of critical social and political analysis, many of these studies have offered important contributions to our understanding of the relationship between violence, war, and life in the early twenty-first century. At the same time, more sustained reflections on the relationship between biopolitics and thanatopolitics derived from Foucault's work, but also willing to raise possibilities not entertained by Foucault himself, have introduced new debates (and problems) within the so-called domain of biopolitical analysis. More importantly perhaps, studies by Agamben, Esposito, and Dillon and Reid have revealed that the referent object of so-called biopolitical frames of representability has changed since the moment when Foucault stated that the object of biopower is "living man, man-as-living-being."[63] While biopolitical frames of representability and intelligibility still generally revolve around the notion of life, the possibility that the promotion of life, the creation of death worlds, the management of dead life, the immunization of live organisms, or the maximization of the becoming-dangerous of the human species may all be referent objects of biopower, seem to indicate that the assumption biopolitics offers one critical explanatory framework, or one analytical perspective, cannot be completely supported. And if biopolitics is characterized not so much by adherence to one Foucault-inspired model, but rather by a plurality of perspectives (or frames of representation of life/the body) that radiate, sometimes in very different directions, from an initial Foucaultian moment of analysis, then the utility of the label biopolitics may need to be questioned too. As we saw above, that label itself was the subject of several interrogations by scholars who thought that additional terms (necropolitics, thanatopolitics, or, perhaps, immuno-politics) were better able to capture the analytical challenge posed by the focus of power on life. More than a label, a model, or a recognizable frame of representability, biopolitics may perhaps better be understood as a general analytical perspective inside which various critical approaches to the interplay between power, war, violence, and life share a family resemblance.

Thinking war, violence, and life beyond biopolitics

The critical perspective on war, power, violence, and life developed in this volume also shares this family resemblance with many of the biopolitical insights introduced above. But while we do not intend to operate an abrupt rupture with biopolitical approaches and what they bring into focus, we believe that, in a manner introduced and supported by Butler, we also need to inquire into what is left outside of biopolitics' frame of vision and representation. While we recognize that the study of biopolitics does "bring into view" (to use Rose's terminology[64]) a series of crucial issues and problems about politics, violence, and life, we must also ask whether some perspectives do not remain out of sight, kept outside the biopolitical frames of representability (whether those are photographic, as with van Agtmael's war images, or textual and theoretical, as with the critical philosophies introduced in the previous section). Thus, it is in this sense of looking at or inquiring about the relationship between war, power, violence, and life outside or *beyond* the biopolitical frame of representability that the title of our book, *Beyond Biopolitics*, can be understood. By mobilizing the term "beyond" here, we do not aim to move past biopolitical approaches, to overcome them, or to introduce totally new analytical perspectives where biopolitical approaches once stood. Our point, rather, is to interrogate what it is that biopolitical perspectives leave unseen or unsaid at the very moment when, or through the very act of writing and representation by which, they present their powerful analytical insights. By departing from biopolitical approaches and trying to think beyond biopolitics, we seek to reveal some of the contextual analytical conditions of possibility that allow contemporary biopolitical perspectives on the war on terror, political violence, the politics of fear, death and enmity, and so on to make critical sense. In the process, and as we will clarify below, by looking beyond the biopolitical frames of critical analysis and understanding, we also introduce, or better yet, re-introduce or re-mobilize, a series of additional (sometimes complementary, sometimes contradictory vis-à-vis biopolitics) theoretical concepts and political positions, which may have been kept out of sight in contemporary biopolitical studies.

However, in order to make our point clearer, it is also necessary to add that looking beyond the biopolitical frame of critical intelligibility does not mean that we must discover analytical perspectives or material political substance to work with that no other scholar (biopolitical or otherwise) has discovered or worked with before. Rather, looking *beyond* biopolitics first and foremost requires us to look *into* the biopolitical framework, and to determine what biopolitical arguments are and how they are presented. This posture is, we believe, precisely what several theorists such as Agamben, Esposito, and Dillon tried to do too when they chose to build from and operate a shift with regards to Foucault's analytics of biopower. Thus, in a similar spirit of theoretical conversation and critical engagement as that displayed by those scholars vis-à-vis Foucault, we wish to advance concepts and remobilize certain perspectives in a way that maintains a creative dialogue with some of the theories of biopolitics introduced in the previous section. Again, it must be reiterated that our invitation to look beyond biopolitics is not a call to surmount or replace biopolitical perspectives. As much as we want to interrogate the conditions of possibility of the biopolitical frame of critical thinking, we also realize that we remain indebted to those perspectives (to what is inside the frame). Thus, to repeat, our approaches and frames of analysis throughout this study do bear a family resemblance with the biopolitical genre, perhaps inevitably so.

One of our analytical points of departure for this study (and one of our points of critical engagement with biopolitics) is the issue of the violence that is being done onto life, the body, and humanity. When reflecting on contemporary instances of violence, Adriana Cavarero claims that "[a]s violence spreads and assumes unheard-of forms, it becomes difficult to name in contemporary language."[65] Biopolitical analyses, we would like to suggest, are precisely attempts to name or give a language (or to provide a certain critical frame of representation) to contemporary violence. Indeed, with biopolitical perspectives (or visual frames, as seen above), it becomes far less difficult to label, explain, or analyze contemporary violence. This is of course one of biopolitics' greatest strengths, and biopolitical frameworks are to be commanded for explaining ("bringing into view," once again) what often seems to make little sense, or, rather, what makes sense, but only after those who mobilize the power to kill on behalf of life have provided their own justifications. But, as Butler argued, making sense of or providing a narrative for what is going on always involves the operation of a frame. Biopolitical explanations, inevitably perhaps, provide a series of critical analytical perspectives that have already been framed by adherence to a previous model, one often inspired by Foucault's analytics of biopower. Among the biopolitical explanations that name and seek to make analytical sense are those that connect, generally via Foucault, terror to life through war and its contemporary violence.

One of the better known biopolitical analyses that connects life to the war on terror is Michael Hardt and Antonio Negri's *Multitude*,[66] the sequel to their path-breaking first contribution *Empire*. In *Multitude*, Hardt and Negri argue that war has become "the primary organizing principle of society" precisely because war has turned biopolitical.[67] To claim that war has become biopolitical or, as they put it, is a "regime of biopower" signifies that war, and particularly the war on terror, is "a form of rule aimed not only at controlling the population, but producing and reproducing all aspects of social life."[68] A war that, as Hardt and Negri suggest, must permanently "create and maintain social order" and promote life can never end.[69] Indeed, such a war must be constant and continuous since its very objective is the preservation of life itself, a life said to be threatened by war and violence likely to be perpetrated by forces that reject a certain (Western, liberal) way of life. A complementary exposition of the connection between war, terror, and life is provided by Julian Reid who argues that "the major significance of [the war on terror] is that it represents the culminating point in the development of a form of regime which has aimed, from its outset, at the subjection of the life of its societies to principles of organization deriving directly from its own war requirements."[70] In other words, in order to preserve life from the disasters and atrocities of war, liberal regimes have to conduct wars (or pacification campaigns) by deploying techniques and strategies that must target not only the lives of those others or enemies that have to be destroyed in the name of life promotion (and those others are often presented as being racially different, as we saw above), but also the lives of populations in need of preservation, protection, reinforcement, or perhaps reinvention. In a typically biopolitical fashion, Reid believes that "the liberal desire to save human life from its subjection to the condition of war … [is a] terrorizing project which can only proceed on the basis of the most resentful violence against life."[71] The endless defense by means of war of the vital conditions of the species or humanity is at the heart of the liberal project and its biopolitical operations of global pacification (wars to end all wars). In this analytical context, the current war on terror launched by the liberally pacifying United States after 9/11 is no different from all other liberal wars, wars being waged on life but also on behalf of life itself. Or, as Reid concludes, "the name given to the defence of that vitality today is Terror."[72]

Looking at the war on terror in this manner does not yield many unexpected results. Indeed, the examinations of the relationship between war and life offered by Hardt and Negri and Reid fit well within a biopolitical frame of representation and understanding. These studies fit the biopolitical model for two reasons. First, they place the war on terror in the context of a larger perspective (itself infused with a biopolitical logic) on the historical liberal way of war or liberalism's constant urge to pacify. Within this historically established frame of analysis, the war on terror is but one particularly violent or ferocious instance of the long liberal quest to wage war on life and to subject populations to "the organizational needs of regimes" obsessed with achieving "increased efficiency in [endless] preparation for war."[73] Second, these studies make perfect biopolitical sense because they subsume the logic of terror under a logic of war. Put differently, terror in a contemporary biopolitical frame becomes synonymous with war and its violence, since terror is the name given both to the war that threatens humanity's (liberal) way of life and to the war that seeks to promote it. Thus, what we are left with as a result of biopolitical perspectives on war, terror, and life is a series of circular arguments whose final referents are the principles of equivalence between war, terror, and life imposed within the biopolitical analyses themselves (war is terror, terror is war, terror is life, life is terror, life is war, and war is life). The self-referentiality and self-sufficiency of the biopolitical frame, and its corollary occlusion of what might reside beyond this frame, explain, for example, why terror itself (what it might mean, why it is rhetorically mobilized, why it is preferred in discourses of war today to other forms of violence or destruction, and so on) is never carefully considered or studied in many critical analyses that look at contemporary violence, in war and beyond, through a biopolitical lens. Instead, in these analyses, terror typically is taken for granted, and terror as war (and the war on terror) is said to represent a wide array of techniques and strategies that, typically, seek to take control of life.

To consider terror (independently or in relation to war) and other modalities of destructive violence, we may have to step out of the biopolitical frames of representation, vision, and intelligibility, as we intimated above. We may have to look beyond biopolitics. One recent study that prompts us to look at terror, violence, war, and life beyond biopolitics is Cavarero's original revisiting of the concept of horror. In her book *Horrorism*, Cavarero suggests that the complexity of the notion of terror may be better apprehended through terror's opposition to horror. Further, she intimates that contemporary instances of violence, destruction, and death—particularly when closely tied to human lives and bodies—may better be thought within the context of horror rather than that of terror (although her purpose is not to replace terror with horror as a preferred paradigm). Terror, Cavarero claims, compels lives and bodies to be mobile and to flee. As Cavarero puts it: "[a]cting directly on them, terror moves bodies, drives them into motion."[74] According to Cavarero, what terrorized lives and bodies seek to escape is a source of terror or a fear that has taken hold of "a certain ordered disposition."[75] With regards to this "ordered disposition," terror creates an instance of mobility, one that can be sudden, abrupt, unexpected, and perhaps unexplainable.[76] Terror pushes lives and bodies to move in order to try to stay alive. By contrast, horror freezes and stops the body or human action. Whereas terror puts one in motion, horror "denotes primarily a state of paralysis."[77] To horrify is to be in the presence of "a scene [that is] unbearable to look at" and that provokes repugnance or disgust.[78] As Cavarero notes, horror (far more than terror) is about vision. Horror is a matter of what one cannot escape, one is forced to witness, and that haunts one's mind, psyche, and body. With horror, the theme of survivability or desperate protection of the body/life (through motion/mobility) is no longer present. Instead, Cavarero goes on to argue, horror is beyond survival and life. Horror is about the visual, physical, and visceral loss or rendering meaningless of all humanity.

In scenes of horror as opposed to instances of terror, biopolitical motifs lose quite a bit of their relevance. Indeed, mobilizing a power of or over life is beyond the point. Instead:

> the physics of horror has nothing to do with the instinctive reaction to the threat of death. It has rather to do with the instinctive disgust for a violence that, not content merely to kill because killing would be too little, aims to destroy the uniqueness of the body, tearing at its constitutive vulnerability. *What is at stake is not the end of human life*, but the human condition itself, as incarnated in the singularity of vulnerable bodies.[79]

By stating that horror (and its performance) targets not so much the "end of human life, but the human condition itself," Cavarero implies that the objectives or effects of horror, differently from terror perhaps, are not apprehensible within a biopolitical perspective or, as she puts it, in relation "to the questions of '*bios*' and 'bare life'."[80] Horror is not reducible to the central theme of biopolitical analyses, that of the need to take command of life in order to preserve it from situations that may weaken it to the point of death. Rather, beyond the concern over life and its vital processes, horror operates at the level of the "savaging of the body as body," or of its destruction as a "figural unity,"[81] irrespective of whether this body is alive or not, or of whether it is a body that is classifiable as *zoē* (bare life) or *bios* (qualified, political life). Here, we are reminded of some of the reflections Primo Levi offered on the kind of horrific violence and hatred that was encountered in the Nazi concentration camps. Levi argued that the typical *modus operandi* of the Lager was not just an endless preparation or readiness to put to death (or a necropolitics of the concentration camp, perhaps), but rather a multiplication of tasks, techniques, and tactics that would demonstrate that "the Jews, the Gypsies, or the Slavs [were] nothing but cattle, mud, and garbage."[82] Thus, the words and deeds that filled the "life" of the camp were, as Levi puts it, "non-human, or better yet, anti-human."[83] The violence or brutality that was on display in the Lager was reflective of a "hatred foreign to humankind."[84]

Beyond life and beyond death, the scene of horror works on individuality and singularity, on the idea of the one or unique human essence (or the essence of the one), on the *ontos* as fundamental presence or as an "existing thing" (in this case, human existence).[85] Horror works on the *ontos* (or performs an ontological function, Cavarero suggests) by dismantling and pulverizing it, that is to say, by breaking up the subject and his/her body into pieces, with limbs torn apart and facial features rendered unrecognizable and undistinguishable from those of beasts or non-living things. Thus, horror is an "offense to ontological dignity,"[86] a dignity that typically presents itself as or is maintained through an insistence on singularity, individuality, and uniqueness of being.[87] With horror, human agency is "distributed across an ontologically heterogeneous field," as Jane Bennett puts it, instead of "being a capacity localized in a human body or in a collective produced (only) by human efforts."[88] The ontological dignity attached to the human body, Cavarero affirms, is a principle that cuts across life and death. Ontological dignity is thus revealed as being beyond life and death, but only at the very moment that the horrific scene enters our field of vision and introduces a violence or bestiality that shows total disdain for such a unity of being.

Such a perspective on horror, or on what is beyond the body or the human as a matter of maximized life or minimized death (or maximized deaths that allow some lives to thrive), is typically invisible in biopolitical frames of representability. Biopolitical frameworks cannot fully capture the horrors present in contemporary wars and many other instances of geo- or bio-political violence (such as, for example, those that Mbembe enumerates in his work) because biopolitics and biopolitical analyses essentialize life or, rather, primordialize a certain form of life that they take to be biological or natural living (or what Agamben calls "simple natural life," once again). Even when biopolitical perspectives detail how the concern with the promotion or management of natural life is transposed from the level of the individual or the population to that of the species in its entirety, these analyses retain a fundamental emphasis on the maintenance and reproduction of techniques and strategies that either enable or disallow the processes of living, acting, moving, and functioning of a live body, a body that must be alive (even if it has no political value), at least potentially so, in order to remain useful to operations of power, government, and force.

Looking beyond biopolitics or outside the biopolitical frame, as Cavarero does through her study of horror, does not allow us to highlight the importance of operations of power that target the vital processes of natural (or, better yet, re-naturalized) life. Nor does moving beyond biopolitics authorize us to restore an analytical focus on a qualified form of life or humanity, on a humanity whose ontological dignity or truly human essence would be in need of salvation or rescue (and, at times, Cavarero's analysis, similar perhaps to some of Hannah Arendt's writings on the traumas faced by the human condition,[89] comes close to intimating such a need for the restoration of ontological dignity). What exploring conditions of terror and horror outside or beyond biopolitical frames of representation enables us to do, rather, is to realize that there are forms of violence on and destruction of lives and bodies that are atrocious—and unnamable perhaps (though not invisible)—because they have moved beyond the threshold of both life and death,[90] or because the condition of being alive or being dead is no longer meaningful to certain modalities of killing and destroying.[91] Although we should probably heed Cavarero's warning about the difficulty of trying to name this violence onto the human, one name we have chosen to give in this book to this modality of destructive or dismantling power/force is that of "agonal sovereignty."

Agonal sovereignty and the dismantling of lives and bodies

What we call "agonal sovereignty" is a modality of power/violence/horror (or a form of power and violence that embraces horror) within which the traditional modern or biopolitical distinctions between life and death, preserving human bodies and sacrificing them, or law-creating and law-maintaining order, on the one hand, and unbounded violence, on the other, are no longer operative. Within agonal sovereignty, the dominant logic is neither about protecting individual or collective bodies, nor about the opposite principle, that of planning and carrying out the elimination of certain forms of life. Rather, agonal sovereignty, as we theorize it in this book, evokes the real (or virtual) possibility of a pulverization of these two prior political logics and, in fact, reveals their indistinctiveness in the face of endless scenes of immanent terror, emergent violence, and unsaturated horror. Agonal sovereignty, we argue, is a configuration of power and force that corresponds to the manifestation of horror and its gruesome violence, as theorized by Cavarero.

In the context of agonal sovereignty, objects of power (to the extent that it may even be said that agonal sovereignty has objectives or objects) are no longer human subjects, citizens, selves (as was the case with modern power), or even life, humanity, or the species (the identified objects/objectives of biopolitical power). Rather, what agonal sovereignty targets are body parts, fragments of lives, bits and pieces of singular human experience and flesh and their always readily reopened wounds or scars. Once again, it is the individuality, singularity, or uniqueness of the human experience, condition, or essence, in a word, of being, that agonal sovereignty wishes fiercely to confront and attack. Agonal sovereignty's horror strives to dismantle the human condition and the logic of the *ontos* by irrespectively annihilating the physical, psychical, moral, and mental attributes of human life. Thus, horror indiscriminately mutilates bodies as much as it arbitrarily traumatizes minds and psyches. Horror operates across cognitive and sensitive registers.

Taking our cue from Cavarero's refiguring of contemporary terror and violence through horror, we intimate that, beyond biopolitical frames of analysis and vision, bodies and lives today often come face to face with sights of human dismantling that may not be apprehensible even with the assistance of some of the most recent critical analytical perspectives on power, life, and death. Put differently, with agonal sovereignty, we arrive at a point where biopolitics and biopower (even as these concepts are revisited and redeployed in some of the most prescient analyses by Dillon, Esposito, or Agamben) give way to a traumatic or disastrous version of human ontology, or, better yet, to an ontopolitics where the very meaning and place of being can never be assured and are always threatened.[92] With agonal sovereignty, we find ourselves in the presence—actual or potential—of a sustained, though not necessarily planned or organized, pulverization of being or, as both Cavarero and Arendt would put it, a devastation of the human condition, a human condition that disintegrates into multiple shreds of individuality/singularity reduced to infinitesimally minute bits of fleshiness (thus preventing any sense of a return to a singular human condition).[93] This is, once again, what Cavarero believes is fundamentally at stake in configurations of horror. This is also why horror, ontologically and politically, cannot normally be accommodated within modern frames of representation and intelligibility. Agonal sovereignty, then, may best be understood as a mode of ontopolitical action (or as an ontopolitical tactic)—not just a modality of interpretation—that accompanies horror, follows it in its tracks, takes advantage of it, and provides a general machinery for the dismantling or deconstruction of human lives and bodies.[94] As such, agonal sovereignty may appear to start with biopolitical elements and may seem to accept the premise of the importance of life in political designs. But it also seeks to transmute the biopolitical into an ultra-negative ontology of being or of the human that, perhaps, not even the concept of terror much in vogue in biopolitical analyses today satisfactorily can capture.

Despite the fact that horrific scenes of violence, terror, and war take us beyond both modern and biopolitical frames of representability, we still insist on using the term "sovereignty." We do so not because we want to suggest that either centralized state power or capillary instances of biopolitical governance continue to be pillars of authority, order, control, force, or violence in contemporary configurations of horror. Rather, we retain the term "sovereignty" here because, as we show in the following chapters, traces of classically modern and sovereign, biopolitical, or even immunitary power systems and apparatuses still find ways of grafting themselves onto modalities of pulverization or devastation of lives and bodies. Still, the sovereignty that we call agonal in the context of this study (such as the one that we attribute to war, warriors, and the war machine in Chapter 1, and that we qualify as a free and boundless power to reshape the contours of the political order through limitless means of terror) and that we attach to contemporary configurations of horrific violence is a sovereignty that is hardly reconcilable with traditional sovereign structures of power, or with the state system. Here, we adopt a way of thinking that may resonate with the analyses of power and force conducted by Gilles Deleuze and Felix Guattari and, in particular, with their opposition between the "striated space" of the state apparatus and the "smooth space" of what they call the war machine.[95] Not only reducible to the act/practice of war and war-making, Deleuze and Guattari's war machine can be read as a general modality of power, mastery, and indeed sovereignty. It is sovereignty in the sense of a majestic and unbridled destabilizing as well as overwhelming force, but one that is not attributable anymore, and indeed may run counter, to traditionally modern forms of political authority and order. Agonal sovereignty, we want to suggest, operates in a register that reminds us of Deleuze and Guattari's war machine.[96] However, agonal sovereignty also forces us to reassess regimes of power and sovereignty that, some biopolitical perspectives have suggested, supposedly have followed and been adapted to the shift from centralized sovereign power or authority to biopower (or a biopolitical sovereignty, as some have argued).[97]

More puzzling perhaps is our decision to mobilize the term "agonal" in this context. We employ the term "agonal" to suggest the horror that at once results from this ontopolitics of shattered life and evokes a new ontological moment, when not even biopower suffices to provide critical meaning, is revealed at the very instant when a certain combativeness, competition, conflict, and war is gruesomely displayed. Here, we understand "agonal" in a manner that Arendt introduced and Arendtian scholars developed (even if Arendt herself did not wish to push the agonal moment or agony too far in that direction),[98] as the qualifying characteristic of an action that is marked by its combative or competitive engagement. Agonal sovereignty, then, denotes a form of confrontation, an antagonism that, for good or for bad, pits bodies against one another or bodies against structures (or apparatuses, or indeed machines—including war machines), often in a violent and brutal fashion, but, in apparent contrast to Arendt's intended meaning of agony, without any hope for the reconstruction or solidification of a political community. On the contrary, agony or agonal action, in the context of our study, is characterized by a destruction of political orders and a negation of humanity. While the way we understand agonal sovereignty is once again not the primary meaning of agony that Arendt typically would want to emphasize, we nonetheless show in this book (particularly in Chapter 1) that the language of agony or agonal action, despite Arendt's wishes, is always already open to the possibility of an "abyssal act of violence,"[99] and, in particular, a violence against not just what it means to be alive (or dead), but also against what it means to be human.

Beyond Biopolitics explores, details, and ultimately problematizes this agonally sovereign modality of action, violence, negative humanity, and pulverization of lives and bodies. It does so by attempting to show how, both starting with and going beyond biopolitical arguments about the preservation of life and the corollary possibility of death, a certain onto-politics of devastation of being—a certain horror, as Cavarero would put it—reinvests and reshapes traditional modern (and at times biopolitical and late-modern) political categories and concepts, particularly as they relate to the production and effects of violence. Among some crucial modern political concepts and categories of thought revisited by agonal sovereignty and horror (and by this study) are war, war violence, and war violence's relation to political sovereignty (Chapter 1), the relationship between fear and political power (Chapter 2), the spatiality of modern and late-modern politics of exception and exclusion of human bodies (Chapter 3), and the role and place of the idea of the enemy in contemporary geopolitical constructs (Chapter 4).

This book unfolds as a series of interventions that, starting with our conceptualization of agonal sovereignty further detailed in Chapter 1, tackle crucial dimensions of the modern and contemporary global political scene of violence/terror/horror inflicted on human bodies and lives, a violence/terror/horror that, through such a targeting of bodies and lives, seeks to attack the very meaning of humanity. Moreover, to demonstrate the critical purchase of our analytical perspective that we place beyond or outside biopolitical frames of representability, we relate the examination of agonal sovereignty and horror (and their conceptual extensions) in the various chapters to pressing issues of everyday domestic and international politics. This is why contemporary "hot topics" such as the politics of exception, the linkage between war, terror, and territoriality, the liberal way of war, or the construction of radical evil in the context of the Global War on Terror (topics often studied by biopolitical analyses too, as we saw above) remain at the forefront of our analyses.[100] In *Beyond Biopolitics*, the strategy we select both to develop our critical arguments and to contrast them to biopolitical perspectives starts with political theorists/philosophers of the biopolitical condition (theorists like—in no particular order—Carl Schmitt, Giorgio Agamben, Michel Foucault, Hannah Arendt, Roberto Esposito, Judith Butler, Gilles Deleuze, Julia Kristeva, and others) and their texts and, later, moves on to put these thoughts in contact with currently salient issues and topics (such as those mentioned above). These theorists and texts, not coincidentally, are often the same authors and literatures that recent biopolitical approaches in political theory, international relations theory, and critical geopolitics have turned to. Indeed, as we already noted, starting to think or see beyond biopolitical frameworks requires that we understand what biopolitical analyses both include and exclude, or enable and disallow, through their use of some of the theorists and philosophies listed above.

However, this strategy of writing and presentation of our arguments in the following chapters does not mean that practical political issues, situations, or controversies are irrelevant to us. To the contrary, we want to suggest that thinking beyond biopolitical frames of representation and intelligibility signifies that it is impossible to assume, before the critical analysis has taken place, that a given political practice or concept will be able to fit neatly within a certain analytical category. We believe that some (sometimes quite popular) biopolitical studies have had a tendency to do the opposite and assume that "reality" can fit into a previously determined, if not always clearly specified, analytical frame. As Butler has indicated, a fateful consequence of developing critical perspectives within a given visual and conceptual frame is that what is out of the frame is never considered to be worthy of attention or interest (as the operations of the frame are often meant to go without saying). Another deplorable consequence of this tendency is that not only is critical political thinking diminished, but also new or different critical political practices are impoverished. This last point suggests that the possibility to devise both analytical and practical challenges—tentative and always uncertain as these may be—to operations of violence, war, terror, and horror will remain limited as long as we insist on replicating (or fitting in) dominant analytical systems, no matter how critical, innovative, or radical they claim to be. Thus, one of the key objectives of *Beyond Biopolitics* is to provide additional perspectives for critical thinking, acting, and perhaps living/being, by turning to biopolitical perspectives as points of departure for the contemporary political engagement of modalities of terror, horror, and agony, but not as foregone conclusions.

1 Agonal sovereignty

Rethinking war and politics in an age of terror

> As an ever present possibility [war] is the leading presupposition which determines in a characteristic way human action and thinking and thereby creates a specifically political behavior.
>
> Carl Schmitt[1]

Introduction

The hegemonic, hyper-powerful, and perhaps imperialistic politics of the United States since 9/11 has contributed a great deal to bringing international relations analysts, geopolitical specialists, political theorists, and social thinkers in general to reconsider the meaning of sovereignty and its consequence for the modern state.[2] Many of these scholars have suggested that, in the context of the Global War on Terror, traditional juridical, legal, and institutional theories that explain the relationship between dominant and dominated political subjects derived from the concept of modern sovereign power are no longer capable of making sense of what has happened in international affairs over the past 15 years or so. As we saw in the "Introduction" to this volume, proponents of biopolitical perspectives have claimed that a *power over life* has taken command of both domestic and international political relations. In the wake of certain strategies, policies, and activities deployed by governmental agents (warriors, police forces, intelligence services, airport security officers, and so on), particularly in the West, a series of post-9/11 practices of prevention of danger, techniques of management of anticipated crises, and generalized modalities of organization of everyday life in and across societies have been concerned with establishing a system of sovereignty adapted to the needs of biopower. This biopolitical form of sovereignty wishes to produce, regulate, rationalize, and make efficient use of life. It is only secondarily interested in maintaining or reproducing more typically modern centralized modalities of legal, constitutional, and institutional order (a matter we shall return to in Chapter 2). Of course, as we explained in the "Introduction," the discovery of regimes of biopower crucial to the inner workings of sovereignty and the state today is really not new. Much of Michel Foucault's work on the deployment of rational methods and procedures of measurement, calculation, and organization of sexuality in eighteenth- and early nineteenth-century Europe was already concerned with the emergence of this allegedly new regime of power.[3]

To make sense of the effects of biopower or biopolitical techniques on sovereignty in the post-9/11 era, some contemporary political theorists have not just revived Foucault's insights and tried to apply them to the realm of today's global imperial policies and practices of control and governance of terror, often by means of war (though, often, they have done so).[4] They have also found it useful to resurrect Carl Schmitt's discourse on sovereign authority as the sole prerogative of "he who decides on the exception."[5] While at first glance the pairing of Schmitt's decisionistic centralism with biopolitical perspectives may seem contradictory, according to these theorists, Foucault's reflections on biopolitics and the governmentalization of the living and Schmitt's political theoretical claims about sovereignty as a moment of authoritative decision marked by the identification of exceptional situations can and, in fact, do come together. Together, they supposedly give rise to a new modality of sovereignty, one that, once again, might be referred to as biopolitical sovereignty.[6]

In this chapter, we show that what some have started to call biopolitical sovereignty cannot be a sufficient critical perspective. Indeed, biopolitical sovereignty as an analytical framework comes short when trying to capture the specific context of the post-9/11 condition, a condition characterized by an increased exploitation of bodies, control over life, and abusive unpredictable violence. It is a condition that, as we hinted at in the "Introduction" to this book, is not simply marked by a violence-spreading terror but also bears witness to a terror that is never really a final political objective. Indeed, we argue that the terror of the contemporary global condition is seeking to establish or lead the way for a paralyzing and dehumanizing horror (to the point that the very meaning of what it means to be human is at stake, as we argued in the "Introduction"). However, the point where many scholars end their analysis is the discovery of regimes of biopolitical sovereignty (and an identification of many sites throughout the global condition where biopolitics is at work). But this suggested end point is precisely where the critical investigation, disquieting as it may be, needs to start. Thus, as we stated above, the idea of a biopolitics of sovereignty can only be a preliminary conceptual platform from which other analyses (including ours) must depart. To make sense of the current condition of global violence and to the mechanisms of horror this condition enables, we agree with proponents of a biopolitical sovereignty approach that several insights from both Schmitt and Foucault can be retrieved. But, unlike what some biopolitical sovereignty scholars do, we do not believe that both systems of thought about sovereignty and power should be fused so as to arrive at a satisfactory or even comprehensive explanatory model. Instead, we argue that singular traits from each theorist's thought can be revisited without having to decide whether they are complementary of each other or whether they can give rise to a new analytical approach to sovereign power and force. Some of these individual characteristics—for example, Schmitt's emphasis on the political as a domain of conflict and struggle, or Foucault's insistence on the way discursive practices are productive of subjectivities—are important insights that help us to be on the lookout for signs of a different modality of power, force, and violence, one that is not simply biopolitical (although, at some level, including its involvement with life, it is), but rather thrives in and indeed relies upon a general context of total warfare and its endless aftermath (that is to say, a context where brutally violent, destructive, and limitless military actions or martial structures of organization of society often deployed by politically disinterested agents ensures that an endless possibility of combat, fear, and horrific dehumanization will take precedence over any notion of political success). In this chapter, following the analytical path traced in the "Introduction," we call this emerging modality of sovereign force and horror *agonal sovereignty*. To repeat, we call it "agonal" because it is in the agony (both in the sense of competition *and* heroism, of a purposeful *and* aesthetic action) of combat and fighting that this sovereign violence leading up to terror and horror manifests itself.

To make sense of what we believe can be referred to as agonal sovereignty, we also take a detour through Hannah Arendt's thought on political action. In the geopolitical context of the past 15 years or so, as political power and the war machine often have become indistinguishable from one another, what some Arendtian scholars (Seyla Benhabib, in particular) argued were two distinct (even, if possible, complementary) modalities of politics for Arendt—the politics of agonistic or agonal engagement on the one hand, and the politics of narration or deliberation on the other—have become increasingly (and perhaps tragically) blurred. Narratives and discourses of war, enmity, or destruction of the other have been precisely—and terrifyingly—what has enabled the deployment of policies and strategies (at home and abroad) geared towards utter dehumanization, the defacing of the other, and horror. Conversely, the boundless fighting, struggling, and fierce competitiveness of the post-9/11 "agonal" warrior (heroic as his or her performance may have been deemed to be at times) is what has further justified the multiplication throughout the global polity of discourses and narratives about the need to reinforce sovereign power and force. Thus, we take an analytical detour through Arendt's work on politics and action *not* to apply her theory of the political (and her supposed search for a political space freed from violence) to the current condition of generalized war, terror, and horror. We shall leave it to others to do this. Rather, we turn to Arendt to garner from her a conceptual vocabulary or language about politics, power, and agony, a vocabulary/language that we adapt to our own analytical purposes. This is why and how we deploy the term "agonal" in relation to sovereignty, politics, or power rather than the more traditionally Arendtian word "agonistic." To further clarify a point first made in the "Introduction" to the volume, by employing the term "agonal," we wish to indicate that it is not a particular philosophical tradition that we seek to retrieve from Arendt but, rather, a certain form of language about engagement, combativeness, and possibly violent struggle. While this language may not always give rise to a politics of total destruction or dehumanizing horror (as Arendt herself hoped to show), there is also no guarantee—other than by placing certain constraints or preconditions with regards to the production of such a language or vocabulary—that it will not do so.

Schmitt, Foucault, and biopolitical sovereignty

Theorists interested in establishing a connection between the state of exception and sovereignty to provide accounts of the politics of war, terror, and limitless power displayed by some states (the United States above all) and their agents since 9/11 have often been tempted to revive Schmitt's understanding of sovereignty. Schmitt's famous statement, "Sovereign is he who decides on the exception," captures the nature of the relationship between the state of exception and sovereignty in his theory.[7] For Schmitt, the state of exception essentializes the functional power of the sovereign by endowing sovereign authority with the ability to make a "genuine decision." The decision on the exception is not only the moment when the sovereign uses extraordinary powers to suspend the law in order to confront something that represents a quasi-existential danger to the state, and perhaps to better restore the juridical or constitutional order. It is also the mark of sovereign force, or what gives sovereignty its defining characteristic and its overall utility. This essence of sovereignty rooted in exceptional decisionism is in line with Schmitt's understanding of the political. For Schmitt, the basis of politics is an endless configuration of struggle or opposition between friend and enemy. The friend versus enemy confrontation gives the political domain its unique nature. It is what makes politics fundamentally political.

However, some contemporary theorists of the state and the exception have suggested that this understanding of Schmitt's sovereignty is perhaps only half of the story. Schmitt's derivation of sovereignty from an exceptional state and his conceptualization of the political as a constant search for enemies may also be associated with a peculiar juridical understanding of the law, namely, the idea that the law must be preserved, protected, and restored (this has been seen by some Schmittian commentators as a remaining legacy of modern social contract theories in Schmitt even though Schmitt did profess his contempt for liberal democracy, pluralism, and parliamentary rule, which he saw as usurpations of the political). Seen this way, political unity—the unity of the *demos* as Schmitt puts it in an essay on the "ethic of state"—remains crucial to Schmitt's theory of sovereignty.[8] In fact, Chantal Mouffe goes so far as to suggest that the unity of the *demos* is the actual foundation of Schmitt's concept of the state.[9] From this perspective, the need to always restore national unity, to reestablish order when it has been broken, and to return to a so-called normal state of affairs can be perceived to be crucial to Schmitt's political reflections. This interpretation also allows some Schmittian analysts to believe that the sovereign decision to suspend the law is always meant to be exceptional and to remain temporary, and thus it can be reconciled with an idea of democratic politics.

This interpretation of Schmitt's exception is driven by a fundamental paradox. Indeed, what some have referred to as the paradox of Schmitt's sovereignty expresses itself as a contradiction between the use of the state of exception (and its decisionistic violence, its abrupt rupture of the normative order) and the juridical/political order that the unity of the *demos* allegedly depends on and that the sovereign decision claims to restore. As William Connolly suggests, this paradox of sovereignty may not be unique to Schmitt (it may even be key to the aporia of law in democratic politics).[10] In Schmitt's case, however, what is interesting is not so much the paradox itself as the way some contemporary theorists choose to ignore (as was the case with the perspective mentioned above) or even resolve what appears to be an important contradiction at the heart of Schmitt's political thought. One such eminent contemporary theorist who seems to have sought to resolve this Schmittian contradiction is Giorgio Agamben. Calling Schmitt's notion of sovereignty a "borderline concept,"[11] Agamben has turned to the ideas of biopower and bare life to situate Schmitt's sovereign exception at the point of departure of a long historical trajectory of power (from pre-modern times to the contemporary global war on terror) whose ongoing objective has been to take charge of life. Crucial to this reading of Schmitt is the idea that what can reconnect the two seemingly contradictory possibilities present in Schmitt's work (sovereignty as absolute decisionism on the one hand, and sovereignty as a derivative function of the need to preserve a democratic unity on the other) is a power over life. Executive decisions on the exception reveal that this power over life—once again, a sovereign ability to either preserve or destroy life through extraordinary measures as well as to produce a collective, political, and unitary way of life within the state by killing enemies outside that state's borders—was always at stake in Schmitt's thought.[12] In this interpretive fashion, the paradoxes in Schmitt's political theory are smoothed over. Importantly, a historical and theoretical continuum (from early modernity to post-modernity) is also established by Agamben. Moreover, Schmitt's views on sovereign decisionism can be turned into the anchoring point for contemporary examinations of practices and regimes of power, violence, and sovereignty.[13]

To be sure, there is more than one critical theorist of the contemporary political condition who has tried to mobilize a notion of biopolitics connected to sovereignty by turning to Schmitt's views on political power. But Agamben has played a decisive role in molding what may become a new analytical framework. What Achille Mbembe has called a "concatenation of biopower, the state of exception, and the state of siege" has been a guiding thread through several of Agamben's political writings since *Homo Sacer*.[14]

In addition to his attempt at resolving Schmitt's paradox of sovereignty, one of Agamben's other tasks in *Homo Sacer* was to situate Foucault's views on biopower in the context of Schmitt-influenced reflections on sovereign decisionism. Here, Agamben has created yet another analytical continuum or conceptual bridge crucial to contemporary biopolitical perspectives: the passage from the state of exception to regimes of biopolitical governance.

Contrary to Foucault, Agamben does not believe that the emergence of biopower is linked to a shift in practices of power within modernity only, and particularly to the development of new technologies of governance of the self around the turn of the nineteenth century. Agamben de-historicizes Foucault's concept of biopower and affirms the presence of a complicity between the Schmittian and Foucaultian perspectives on power (and by extension on sovereignty). Without de-contextualizing Foucault's analyses of biopolitics, Agamben and others like him who operate within a critical framework of biopolitical sovereignty (Hardt and Negri, for example) would have to deal with the question of how it can be that Schmitt and Foucault actually perceived sovereignty and power differently even though both of them based their views on nineteenth century configurations of order, power, discipline, and force. This is partly the reason why Astrid Deuber-Mankowsky has recently argued that the distinction between Schmitt's decisionism and what she calls Foucault's "nominalism" is crucial and must be maintained. Such a distinction, Deuber-Mankowsky claims, is the result of a "decisive difference between Schmitt's concept of the political … and Foucault's concept of politicization."[15] In other words, while Schmitt seeks to uncover and preserve the essence of the political (that, once again, he finds in the sovereign decision on the exception), Foucault has no such designs and the political for Foucault is a field of possibilities for relations of power, knowledge, and subjectivization (the making of political subjects and their governance). Thus, Schmitt's view on sovereignty remains tied to a conception of power that is necessarily centralized (something we will address in more details in Chapter 2) and dependent upon the persona of an authoritative sovereign or state leader. Foucault's view on sovereignty, by contrast, is fragmented, pluralized, relational, and tied to various techniques of government of the subject and the living spread throughout society. It is true that, as Mouffe notes, both Schmitt and Foucault wish to expose the fallacy of democratic principles of modern government and of the rule of law in liberal systems. They also both note that liberal–democratic rights regimes and norms are trapped inside grids of instrumental control and force (something that Max Weber, before them, mentioned too).[16] However, what each theorist takes to be the source, the form, or the effects of sovereign power is quite distinct. Thus, reconciling these two perspectives around the theme of sovereignty and power means ignoring the fact that Schmitt retains a fundamentally vertical or top-down model of power while Foucault's approach is precisely meant to be a challenge to these hierarchically structured conceptualizations of force, authority, and political control.[17]

To circumvent or overcome these theoretical differences, Agamben first appears to accept Foucault's move away from the juridico-institutional model of power. But Agamben also believes that we need to graft onto it a fundamental distinction that, according to him, is to be found in the determination of or over life and that resides at the core of Western political philosophy. The figure of *homo sacer* (and his/her embodiment of bare life) is crucial to this originary philosophical distinction at the level of life and, furthermore, it is also essential in Agamben's thought to the fusion between Schmitt's and Foucault's views on power that Agamben wishes to promote. *Homo sacer*, or the human body conceived as a function of useful characteristics of productive live forces, but never as a participating political subject essential to the realization of the good life in the *polis*, finds itself in Agamben's work at the junction of Schmitt's sovereign exception and Foucault's practices and regimes of biopower. This "sacred being," or s/he who "can be killed but never sacrificed," occupies a particular zone of indistinction between "exclusion and inclusion, outside and inside, *bios* and *zoe*, right and fact."[18] The political state of exception opens up the possibility of creating a spatial domain where the individual body becomes an object abandoned to the full arbitrary force of sovereign power (more will be said on this in Chapter 3). Or at least, and this is precisely where Foucault's thought is both inserted and expanded upon by Agamben, in this zone of sovereign indistinction where bare life is at play, the full range of regulating, disciplining, but also efficiently functioning techniques of government from a wide array of institutions, agents, and apparatuses is applied to the life of the population, a population or society whose destiny is always said to be at stake in such biopolitical designs. Thus, for Agamben, the sovereign ban and the protection *as* production or subjectivization of a certain way of life are part and parcel of the same philosophical configuration of power and of the same long-term historical process, namely, sovereign power's reliance upon and utilization of bare life.[19]

Agamben's synthesis of Schmittian and Foucaultian notions may be compelling at times and seductive to many critical scholars. Several contemporary chroniclers of the biopolitical condition, starting with Hardt and Negri (in their trilogy *Empire*, *Multitude*, and *Commonwealth*), have willingly accepted it. But it also ignores or conveniently bypasses crucial elements in Foucault's analysis of biopower. Among other things, what Agamben's perspective neglects are the detailed, varied, and discontinuous micro-historical accounts of the ways societies and cultures have internalized juridical, political, and economic relations of power, force, and knowledge (something Foucault's work is attentive to). Instead, Agamben wishes to prioritize a meta-narrative about the relationship between modernity and sovereignty, one that inevitably minimizes the role played by various local, regional, national, and (lately) international discourses of power relations and power effects that, for many individual lives, give sovereign power its meaning and force in concrete contexts. According to Connolly, this desire to impose a historically dominant and supposedly coherent "logic" of sovereignty (biopolitical or otherwise) betrays a will to transcendence in Agamben's philosophical project.[20] Still, it remains the case that, for many "docile bodies" out there, what is remarkable about sovereignty is far less its vertical structure or centralizing capacity in moments of exceptional decision-making than the many disseminated instances of control, knowledge, subjectivization, and rationalization of violence that depend on various discursive productions through which authority and exceptionalism can achieve discrete, efficient, and often coercive and conditioning effects throughout societies and on case-by-case bases too. For many bodies, lives, or populations in today's global polity, sovereign power, or the decision on the exception, is received and perceived as a series of decentered, discontinuous, and disembodied performances of force, violence, terror, and horror that are as discursive as they are material.

To arrive at an apprehension of the way practices of day-to-day living and being (of living as a human and being human) are increasingly captured by regimes of sovereign brutality and violence, we can move away from Agamben's biopolitical framework. We can look beyond biopolitical frames of representability such as those that keep on repeating that war violence has become a dominant biopolitical motif or an expression of sovereign exceptionalism. We need to move towards analytical perspectives and concepts that are not primarily interested in establishing biopolitical sovereignty as an all-encompassing (or all-bridging) explanatory model. As we stated in the "Introduction" to this book, such a move implies that perspectives on biopolitical sovereignty (like those spearheaded by Agamben and reliant on an alleged continuum of critical thinking about sovereign power from Schmitt to Foucault) can instead become the starting point for additional theoretical investigations, in particular for investigations into the meaning and place of violence in an age when terror operates as a gateway for a horrific humanity. In the next section, we turn to Arendt's language about political action to look for a conceptual vocabulary that could facilitate a different critical analysis of the effects of the multiple regimes of sovereignty and power in an age when the horrors of a war that keeps on bearing terror's name continue to dominate public decisions and discourses, and when heroic warring agents are still primarily in charge of preserving the so-called integrity of the body politic or unity of the nation. Put simply, it is through Arendt that a language of agony, agonal action, and agonal sovereignty can be discovered.

Arendt, action, and agony

A conventional reading of Schmitt asserts that the sovereign power of decision is about choices regarding enmity. This claim further implies that it is about choices regarding when, where, and how to go to war. Paul Hirst writes that, according to Schmitt, "it is political struggle which gives rise to political order."[21] What Schmitt and theorists who turn to sovereign exceptionalism as an explanatory model eventually cannot account for are public actions performed by war-makers, warriors, or other agents of conflict, combat, and confrontation *after* the decision to turn to such a war machine has been made. In other words, they do not have a rationale for the kind of necropolitical logic (as Mbembe puts it) or thanatopolitics (as Esposito might call it) that warriors follow, perhaps because such actions are allowed to take place freely in a post-decisionistic domain, where certain agents are neither lawful nor unlawful, neither in command nor under control, neither inside nor outside (of politics, of state borders), and, ultimately, neither sovereign (they did not initiate the decision to go to war) nor subordinate (they still are in charge of determining what tactics will be utilized to destroy the enemy). Thus, a blindside in the biopolitical perspective may be identified here again. Indeed, proponents of the biopolitical sovereignty approach are not able to think past the moment when the decision to launch the warriors on their destructive paths has been made (other than to proclaim, as Agamben and a few others do, that through war the state of exception becomes a permanent condition). Critical political and international relations theories that have turned to biopolitics/biopower of late may not possess the conceptual vocabulary that could problematize *and* make intelligible the relations of force, power, terror, and horror that unfold once war-makers take over the conduct of the political and sovereign domain. Put differently, critical biopolitical analyses are still not equipped with a frame of intelligibility that could capture the sovereign exception as what Slavoj Žižek has called "an abyssal act of violence,"[22] that is to say, as a decision that turns the political into a space of both terror and horror, inside which warriors insert themselves, map out plans of attack, take advantage of decisions made through or, rather, as combat tactics or battle maneuvers, and, in this martial fashion, take command of human life.[23]

Arendt is an interesting critical voice to turn to in this context because, in her work, she did recognize the danger for humanity of the abyssal violence always present in the sovereign decision. The danger of sovereign decisionism, as Arendt saw it, stemmed from the fact that the sovereign/sovereignty was left to operate in a political/power vacuum that nonetheless could be expanded *ad infinitum* and perhaps beyond the sovereign's own reach. As Andreas Kalyvas has argued, Arendt did understand that "decisionistic politics [of the Schmittian kind] was trapped in its own deceptive omnipotence."[24] Kalyvas further claims that "[u]nregulated and faced with no limitation, unshaped and boundless, the sovereign decision became vulnerable to its own transient and fluid dispositions to plunge finally into terror."[25] Related to this pertinent warning with regards to sovereign decisionism, Arendt's reflections on the nature of the political decision are useful to us for another reason. Indeed, as mentioned above, they provide us with some suggestions on how a language of war or even of warrior performance—and a corollary perspective on political relations dramatically reconfigured through heroic deeds of warring agents—could take shape. In what some Arendtian scholars may find to be an unexpected use of her work, Arendt's political theoretical vocabulary, if not the full measure and substance of her analyses, allows us to find an analytical point of entry into the domain of the war machine and of agonal sovereignty.

Many have shown that Arendt's modality of political action can be situated in the context of a public sphere that, according to Arendt, needs to remain plural, free, and, more importantly, contingent. As Lisa Disch notes, Arendt's politics is agonistic because its nature (taking us back to Greek tragedy) is to be aesthetically or performatively competitive and combative within such a setting.[26] The term "agony," it is worth noting, is derived from the Greek word *agon* that designates a public contest between certain actors (agonists) towards the achievement of a prize. Agony (*agonia*) signifies contest or competition, although Webster's dictionary also refers to this contest as a violent one.[27] In a Judeo-Christian context, agony (eventually conceptualized as an inner battle, struggle, or suffering prior to death) is also the moment that can open onto a revelation.[28] It is an ultimate (and, again, sometimes violent) struggle/battle with the self and the body that enables the soul to free itself from the human world. In agony—and it is important to keep this dimension in mind in the context of the relationship between agonal sovereignty and horror—both the body and the condition of humanity are eventually surpassed. The very meaning of leading or being a human life gives way to another transcendental perspective. Thus, in the threshold condition that is agony, there is a fight to surmount or reach beyond that which is corporeal about the human. In a way, in its Judeo-Christian genealogy, agony already announces a desire (expressed through a violence and perhaps horrific struggle) to overcome a politics of life and the body at both an individual and a collective level.

Politics for Arendt is a creative engagement on the part of public agents who, both in competing amongst themselves to realize a remarkable performance in a public context (in the political arena) and in fending off previous structures of power, demonstrate their "greatness." Indeed, Disch intimates, the idea of greatness is crucial to Arendt's understanding and usage of the term "agony," particularly when the term is made synonymous (by Arendt, to start with) with action or performance. Arendt declares that "action can be judged only by the criterion of greatness."[29] Yet, while greatness appears to be crucial to Arendt's agony, greatness itself is not sufficient to understand what agonal action implies. Rather, it is greatness *coupled with* competition, combat, or struggle, that is to say, greatness valued as a daring, creative, and perhaps impulsive engagement in a public setting, that matters here. The qualifying characteristics of being combative, competitive, creative, or daring attached to greatness are not meant to imply that the glory or brilliance contained in Arendt's agony is one that automatically will lead to violence or brutal conflict (although such qualifiers *do not* exclude violence or conflict either). Rather, they are to be understood as tragic, aesthetic, or even artistic performative dimensions of what it means to be agonistic or agonal,[30] that is to say, to act with valor or extraordinary prowess in the public domain.

Arendt's agonal action is supposedly free from any pre-existing rationality. It can only attain its political meaning through its momentary brilliance, as an end in itself, or as a self-contained performance. In an often cited passage (from which the statement about greatness above is taken), Arendt describes the significance of producing a "great" action as follows: "Unlike human behavior … action can be judged only by the criterion of greatness because it is in its nature to break through the commonly accepted and reach into the extraordinary, where whatever is true in common and everyday life *no longer applies* because everything that exists is unique and *sui generis.*"[31] Arendt recognizes that the necessary condition for agonal involvement to manifest itself as something great or glorious is an intensity of purpose or a desire to excel on the part of public agents. An example of this type of contingent pre-eminence is the Greek mythological figure of Achilles, a heroic image that Arendt turns to in some of her texts. The story of Achilles became immortalized for subsequent generations of ancient Greeks, and later for moderns as well, precisely because he was, as Arendt puts it, the "doer of great deeds and the speaker of great words."[32] We remember Achilles because of his heroic capacity to act brilliantly, because of his thrilling virtuosity, but also because of his combative spirit. Arendt wants to show that it is not just Achilles' deeds that matter, extraordinary as these may have been. Indeed, the need for a narrative reference and remembrance of Achilles' heroic acts is crucial too. Through narrative inscription or recollection, the political space can become democratic, free, but also durable (the abode of the "memorable deeds and the names of the memorable men who performed them"[33]).

Arendt seems to want to embed the heroic and agonal performance of the warrior in a form of narrative politics, one that, based on rhetorical representation and remembrance, may actually serve as a guarantee that the political would not just be about the deeds of agonal heroes (it would also be about the stories told about those glorious performances and meritorious events). Thus, Seyla Benhabib maintains that Arendt does present two modalities of action. One modality would be "agonistic" or agonal, that is to say, engaged, valiantly combative, and based mostly on great or brilliant performances by some public agents (but, we must ask, can all public agents truly become Achillean heroes?). The other, a so-called narrative modality of action, is also what Benhabib calls "a process of invention."[34] These two models of action are not antagonistic though. Rather, they are complementary, or better yet, as we suggested above, one (the narrative model) seems to add a dimension to the other (the agonistic model). As Benhabib indicates, for Arendt, "all action, including agonal action, is narratively constituted. The what of our actions and the who of the doer are always identified via a narrative, via the telling of what one does and who one is."[35] According to this reading, great deeds alone, aesthetically glorious and heroic as they may be, may never be meaningful without the element of narrative representation suggested by Arendt. Great agonal actions without a narrative follow-up may not lead to the lasting development of a free polity.

At the same time though, this complementary presence of language, narration, or discursive recording in Arendt's political model may also be used (perhaps by Benhabib, to start with) as a justification to downplay the significance of the performative and self-actualizing dimensions of Arendt's political action or agonal intervention. The uneasiness of some Arendtian commentators with regards to Arendt's agonal politics, and these commentators' corollary desire to privilege Arendt's narrative "process of invention," may have to do with the fact that Arendt's action, as a purely agonal and aesthetic event (one without any predestination or instrumental rationality), may indeed be valued for nothing other than its performance, its contingent happening, the mastery or brilliance of its performers at the precise moment when they undertake the deed and leave it up to a future (aleatory and uncertain) exercise of narration, description, or representation to derive some durable meaning. The Arendtian free political performance, as Bonnie Honig has noted, can be understood as agonal also in the Nietzschean sense of a deed, event, or occurrence from which not only instrumental rationality but also moral judgment has been removed. As Honig clarifies, the performative acts that Arendt focuses on "are not cognitive statements; they are events."[36] Their value resides in the performance itself. Thus, what some Arendtian commentators may have been uncomfortable with (to the point of wanting to insist on linguistic or rhetorical remembrance as a supplementary buffer) is the idea that the aesthetic style of the action may matter as much, and perhaps more, in the constitution of the politically engaged or agonal act as the possible substance or meaning of the deed itself (or, rather, the future narrative representation of the deed's meaning). Dana Villa suggests that "the performance of action in public provides the opportunity for stylization; and stylization, in turn, is the precondition for the kind of reification identity demands and for the transformation of a public life into a memorable narrative or story."[37]

The perceived danger of this interpretation of Arendt's politics of action/agony, according to some scholars like Jürgen Habermas, is that not all the members of the public are capable of such heroic/agonal actions. In fact, it may be that only a "specialized minority" is able to produce political greatness for all to see. Thus, Arendt's politics of action may also relegate the majority of the members of the public to the role of a passive and admiring audience, an audience whose participation in the constitution of the political may not be so essential after all (rather, most of the members of the *polis* may be content with their private occupations and leave to exceptional actors the task of occupying and embellishing the political domain). Whatever form of narration or remembrance that may be demanded by Arendt of the witnessing masses may end up foreclosing any meaningfully deliberative mode of communication, exchange, or negotiation, as some commentators have deplored.[38] An additional concern is that the heroic performance or agonal action Arendt appears partially to celebrate has the potential to descend into a physical form of combative conflict and violence (an agony of war) if the linkage to linguistic memorization is ignored, Arendt's wish to keep politics and violence separate notwithstanding. In this way, the philosophical and political lines of demarcation that Arendt sought to maintain between her views on political action and Schmitt's perspective on sovereign decisionism may not be so clear after all.

It is Arendt's language of agonal action/performance (not her political lexicon about memorization, narrativization, or representation) that may have brought her close to realizing that the great act of the heroic public agent—taking place in a space allegedly freed from sovereign decisionism and valued for its performative manifestation—could not offer any promise that it would not turn into an "abyssal act of violence," one that could be deployed for the sake of an endless violent heroism or to perform a violent action just because it can be done. Still, Arendt refused to entertain the possibility that her agonistic politics could fall prey to the whims of agonal warriors, to Achillean heroes "gone bad" and for whom vigor, brilliance, and extraordinary spirit in battle would have become the only standards for public action. Arendt would not have accepted the possibility that Carl von Clausewitz's famous adage about the relationship between war and politics be reversed (a move that, as we shall see below, Foucault suggests was already part and parcel of modern Western societies from the sixteenth and seventeenth centuries onwards). Incidentally, it is worth noting that Schmitt could not accept such a reversal either, but because for him this would have implied that the sovereign's decision had lost control of the friend versus enemy distinction. Arendt could not conceive of this possible reversal because, in her view, it would have signified that any creative, contingent, revolutionary, or competitive new beginning for the political domain always would have had to be subjected to critical doubt or suspicion. In an ironic twist, Arendt's language of agony reveals how tenuous the boundaries can be between maintaining or, in fact, inventing a political domain and destroying it (as well as others or enemies in the same process) by having recourse to the performative acts of heroic public agents who are left in charge, without any limitation or safeguard, of devising or realizing new political possibilities. As military strategy theorist Alain Joxe has suggested, what one ought to question today is the status of political goals, and the performance of such goals, when "there is no global *political* power [anymore], only a global *military* power (the American army) and [perhaps] a global *economic* power (corporations, the market)."[39] If this is the fate of our contemporary "political" situation, Arendt's language of agony/agonal engagement (her emphasis on the heroics of combative or competitive figures) more than her language of memorization and narrativization may give rise to another way of envisioning politics and sovereign performance. This sovereign performance (that of an Achilles-turned-war machine perhaps) stands beyond any sense of what is useful or expedient for a political community (a population, a society, a body politic) to maximize its living conditions and is closer to what we have called agonal sovereignty.

Agonal sovereignty and the warrior's horrific violence

According to Michael Hardt and Antonio Negri, war has become a permanent social condition among sovereign political entities and human beings, a condition of "exceptional" power and force that is no longer exceptional at all. Following Foucault's lead in *"Society Must Be Defended"*, Hardt and Negri recognize that Clausewitz's famous motto has been inverted and that war now constitutes the political.[40] Or, to use the language of biopolitics, war is the crucial formula deployed by governmental agents to preserve or promote a preferred way of life, one that can be experienced both inside sovereign states' borders and throughout the circuits of Empire (as Hardt and Negri put it). Still, while assuming that Empire supervises and manages the conduct of conducts both inside and outside the state, Hardt and Negri also stress the belief that the permanent biopolitical condition of war is not to be seen as a succession of unregulated or excessive forms of violence. In a sense, for Hardt and Negri, war is still subject to some sort of sovereign rule, or rather, to use their terminology, it is always submitted to Empire's global biopolitical designs. Thus, they intimate, the legitimacy of Empire is derived from the capacity of violent and bellicose agents to create order as a certain (economically, culturally, and politically) globalized way of life and to maintain this global "Empirean" order through a constant confrontation (and often destruction) of monstrous enemies. At the same time, Hardt and Negri do admit that this type of legitimation of global power/sovereignty can become unstable as Empire is constantly confronted with the figure of the "Multitude." Thus, Hardt and Negri add that "the asymmetry and imbalances of power in the world [inherent to Empire] cannot [completely] be absorbed within the new legitimation of imperial power."[41]

Those "asymmetries" or "imbalances" mentioned by Hardt and Negri often become triggers for the kind of agonally combative violence performed by heroic agents that escapes the reach of sovereigns or Empire. In the cracks left open by Empire and the remnants of modern state sovereignty, the war machine "smoothens" (to use the term mobilized by Gilles Deleuze and Felix Guattari, and later borrowed by Hardt and Negri too) the space of the biopolitical sovereign exception and its production of life.[42]

What is also notable in Hardt and Negri's texts (similar, perhaps, to what we saw in Agamben's philosophical explorations) is a particular reading of Foucault's analyses of biopower, a reading that allows them to present a coherent and possibly encompassing model of biopolitical sovereignty. The need for such coherence and for the maintenance of a framework of biopolitical analysis is not that clear and not so urgent if, in fact, we take a different path to appreciate Foucault's writings on the relationship between biopolitics and war. In *"Society Must Be Defended"* and a few other seminars and lectures where Foucault turns to the problem of biopower, his intention is to rethink modalities of power by decoupling them from the juridico-institutional straightjacket that characterizes modern theories of the state.

Foucault has his sight set on historicizing the concept of power and, by extension, of sovereignty. And he intends to do so by deploying three successive and complementary tactics: first, by revealing the intensity of the discursive modalities that surrounded sovereignty's contestations over time; second, by highlighting the transition from the classically modern idea of the state to the bureaucratic and procedural implementation of functioning governmentalized apparatuses and institutions; and third, by documenting operations and processes of force, control, discipline, productivity, rationality, and efficiency that are often the main expressions of political life in the modern age. Moreover, to appreciate the impact (for Foucault, at least) of the possible reversal of Clausewitz's dictum into "politics as war" in modern times, Foucault suggests that we need to pay attention to a series of discursive features about war that render problematic (if not outright impossible) any hope of a clear conceptual distinction between a constitutive or regulative violence on the one hand, and a different, perhaps agonal, violence on the other.

As Andrew Neal has indicated in his analyses of Foucault's reflections on the war–modern power nexus, Foucault does not really want to posit the reversibility of Clausewitz's "war is the continuation of politics" thesis and present it as a "present-day response or alternative to the problem of modern juridical sovereignty."[43] Rather, Foucault's genealogies of power and sovereignty, again in *"Society Must Be Defended"* in particular, are meant "to identify a pre-Clausewitzian discourse of 'politics as war' that Clausewitz later inverted."[44] Foucault indeed makes it clear that this "historical discourse of politics as war"—a discourse of agonal *and* destructively engaged performance that becomes the mark of the political domain—is also a discourse that is based on real-life battles, and privileges the place of the heroic warrior in society (thus, this discourse of war is also not a heuristic exercise as it is, for example, in Hobbes' case where it is meant to evoke an abstract and ahistorical state of nature). As Foucault puts it, it is the modern political order and the law itself that are "born of real battles, victories, massacres, and conquests which can be dated and which have their *horrific heroes.*"[45] This "reality of war" motif is then allowed to be discursively expanded to various political and social domains and institutions where it is encouraged to shape the meaning of modern life. Such a discursive serialization of battles, invasions, and conquests narrated, remembered, immortalized, inscribed, and reinterpreted has no specific objective or end point. Instead, the "historico-political discourse" of war is transformed into what Foucault calls an "epistemic web" that guarantees that the political and social utility of war can always be rediscovered, reinvented, reused, transferred, or transformed.[46]

In this fashion, Foucault's analysis of discourses of war and war's agony, of warriors' heroic but possibly senseless and destructive actions designed to have effects in the public domain and, in fact, to shape and reshape the contours of the political compact (to the point of obliterating political meaning and life) can be seen as the accursed share of Arendt's own thought and lexicon about agonistic action. Put differently, Arendt's language of agonal engagement, a Foucaultian reading allows us to realize, always contains the possibility that what she hoped would remain a non-violent public sphere will actually be turned into a space of abyssal or horrific violence (hence, once again, her inability to accept the reversal of Clausewitz's adage). Among other things, Foucault shows us how Arendt's attempt to keep war's agonal violence and the conduct of modern politics separate is the product of a certain will-to-believe in the power of free human creation and its transmutation into a community of politically engaged individuals. This Arendtian will-to-believe relies on a certain cleansing of the concept and language of agony, a theoretical operation that, in a way, has already and arbitrarily decided that agony or agonal action will never fall prey to the whims of the agents who perform the deeds. As for Foucault, both historically and discursively, the specter of an agonal violence that could endlessly penetrate and possibly conquer the space of the political, and of the maintenance of a certain way of life, has already been realized. In fact, this heroic/agonal warrior violence is part and parcel of political modernity and is embedded in the concept of sovereignty, something that Schmitt (one might suggest) was perhaps closer to capturing than Arendt.

Foucault also illuminates the importance of discursive representations that, unlike the kind of narrative representations and remembrances Arendt was hoping for, establish the primary function of heroic warriors (the glory or excellence of their actions, their victories, and their killings) and infiltrate the practices and policies of modern institutions and rationalities of power. Here, Foucault's analysis of historical discourses of war reconnects us with Mbembe's attempt at thinking the contemporary condition of political violence as a form of biopower/biopolitics that has not only life but also the creation of "death worlds" as its main target. For Mbembe, as we hinted at in the "Introduction" to the volume, it is not so much the Schmittian state of exception that is the main characteristic of today's global biopolitical order (as Hardt and Negri might put it). Rather, it is a "state of siege" that is the distinctive mark of pluralized and decentered regimes of violence that, once again, he labels "necropolitical."[47] As Mbembe specifies, "[t]he state of siege is itself a military institution. It allows a modality of killing that does not distinguish between the external and the internal enemy." As a result, "[e]ntire populations are the target of the sovereign. The besieged villages and towns are sealed off and cut off from the world. Daily life is militarized. Freedom is given to local military commanders to use their discretion as to when and whom to shoot."[48]

In the Global War on Terror era, this dominant theme of "politics as war" (Foucault) or even "politics as death" (Mbembe) is frequently taken for granted and reproduced in foreign policy analyses of the post-9/11 American turn to a certain modality of agonal violence, one that entrusts the war machine or heroic war commanders with complete authority to take charge of social and political life, possibly to the point of its total destruction.

For example, war critic Andrew Bacevich's study of the rise of American militarism, and his related reflections on contemporary US hegemony through war, have revealed that, as he puts it, under a "prevailing national security consensus … mainstream politicians today take as a given that American military supremacy is an unqualified good."[49] Associated with this image is the emergence of cultural and political narratives that present this military and martial take-over of the public sphere as an urgent necessity, one that allegedly aims at reaching total global transformation.[50] As Larry George has noted, this turn to the war machine may be a reflection of the United States' desire to rediscover "the ontopolitically restorative effects of both pharmacotic violence and pharmacotic wars to sustain its patriotic myth."[51] For George, "pharmacotic war" is not a completely new phenomenon. It refers to the belief that war is a purgative drug or medicine and that this drug/medicine can rejuvenate and re-mythologize the nation's society, population, and body politic. In the Global War on Terror, the agony of political action and the desire to always look for ways to restore the ontopolitical status of the nation converge (no matter if the proposed agonal cure turns out to be far worse for the body politic than the diagnosed ailment).

Over the past decade, US war pundit Thomas Barnett's writings have been emblematic discursive supports for such a pharmacotic or agonal war performance. Barnett's infamous plea to the US government to turn to "its" war machine and to launch American warriors on a series of military crusades to reduce the "gap" (as he puts it) in today's globalized world between populations that are globally connected (read liberal, democratic, and free-market oriented) and those "archaic" ones whose resistance to globalization is a danger to the US nation, society, and way of life, is a blatant illustration of this verbalization and reimagination of politics (and political action) as an open-ended succession of war battles, war machine-produced deaths, and glorious murders by heroic warring agents.[52] In Barnett's text—or what he believes is his personal strategic "blueprint" for the Pentagon to reshape global political, social, economic, and cultural life—and in those of many contemporary narrators and textual producers of the violently destructive glory of war and war struggles, the sovereignty of the state (the United States, in particular) gives way to an agonal sovereignty, or, if one prefers, to the sovereign majesty of the war machine. This sovereign mastery and majesty through and, in fact, as war may appear to be concerned with re-establishing the ontopolitical basis of the state or the nation, as George suggests. But, beyond this ontopolitical need, it continuously strives to redefine the contours of global (bio or necro) political life. This agonal sovereignty operates in this manner not simply from the perspective or for the benefit of the American state, the *demos*, or even the president/executive decision-maker/leader (this is why ontopolitical conditions of possibility are transcended), nor even for the sake of the regulation or policing of a transnational multitude. Rather, the agonally heroic war machine performs sovereignty functions and tasks for the purpose of being able to produce more and more warring bodies (making potential warriors of us all, irrespective of whether we are friends or enemies) and more and more instances of combat, violence, death, destruction, terror, and horror against human bodies and lives whose identity does not really matter.

In fact, the objective of this agonally sovereign war machine is not primarily to kill other beings (right here or over there) to maintain or restore a way of life. Its objective, rather, is to establish destructive violence—a violence that once again indiscriminately annihilates humans and has no more respect for life than it has for death—as a regime of terror that opens up onto a horrific humanity. While the war machine does not stand in for state power, the decision of the sovereign, or even Empire, it also does not function as a political agent on behalf of human life or humanity. Instead, in the operations of the heroic war machine, humanity is often reconfigured as what precisely stands in the way of war efforts and agonal violence. This disdain for humanity as a key feature of agonal sovereignty perhaps explains the increased attraction that today's warring agents have for technologies of death or, better yet, for technologies of horror, in the battlefield and beyond (to the extent that we can even talk of a "beyond" of the battlefield). We can think, for example, about the unmanned flying Predator drones that track and kill with little moral or political interference from the human mind or human emotions. Military drones (in the air, soon perhaps on land or in the oceans) are deprived of the human capacity to decide when it is too much violence and when it is not. While they are monitored and steered by a "crew" at some military headquarters thousands of miles away from the targeted site, the drones are meant to follow a computer program that "instructs them" who to track and/or attack, where, and when. With limited sentient abilities (generally associated with human life), the drones have virtually no capacity of reaction, response, or improvisation (including change of course or restraint in attack). They are nonetheless active warring agents with a mission, often a kill and destroy mission. As P.W. Singer explains, immediately after 9/11, as the US military started to actively search for Osama bin Laden in Afghanistan, "the idea … arose to arm the drone by mounting laser-guided Hellfire missiles on the wings." Singer continues: "Since the Predator already could direct missiles at targets with its lasers … the drone would carry its own, instead of having to rely on the kindness of strangers to blow up those below."[53]

Those Predator drones are post-human heroes and agonal agents for whom humanity (both in their operating procedures and in what are their programmed targets) means very little. They are post-human in the sense that they represent "a historical moment in which the decentering of the human [is realized] by its imbrications in technical, medical, informatics, and economic networks."[54] They are also post-human to the extent (often not yet theorized by theorists of post-humanity) that this so-called "technological decentering of the human" facilitates the pulverization of the human flesh since any limit condition determining what it means to be or remain human has been obliterated too. Thus, the terror unleashed by these agonal drones can turn into an instrument of horror as they and other robotized machineries of this kind that war commanders and war designers claim to control from afar have no life or death threshold. Or, as Singer suggests, "unmanslaughter" is now permissible and realizable.[55] In the end, operationally and conceptually, only a pulverization of the flesh (the live or dead flesh), only the maiming of humanity can be achieved by such agents that are purposefully deprived of any human dimension (again, even the war commanders make sure they remain at a distance).[56]

As a result of the fundamental inability of the war machine to determine the conceptual boundaries between politics and war or between humanity and horror, we are perhaps faced with what Neal has called "a politics of collective subjective enmity wedded to a terrifying state machine. The question is not simply one of who is or is not being constructed either as 'the enemy of the state' or 'the enemy of the nation/society/people,' but a frightening union of the two."[57] Neal's argument is partially correct, but it needs an important correction. Indeed, it is not enough anymore to suggest that a terrifying *state* machine is in charge of today's operations of terror. Rather, it is a terrifying *war* machine that is at work in contemporary post-human or, better yet, *anti-human* situations of horror (to use the turn of phrase crafted by Primo Levi about the concentration camps[58]). And for this anti-human war machine, biopolitical considerations—the need to preserve or create a way of life or the decision to put to death—are no longer the main determinants of action.

Conclusion: returning to Schmitt?

The possibility that a horrifying but agonally supreme war machine could advance objectives, deploy strategies/plans of action, and perform unspeakable acts of violent dehumanization beyond biopolitics should also prompt us to revisit the opposition between friend and enemy (and, through it, the meaning of self and other, something we do in Chapter 4) that, as we mentioned above, forms the basis of Schmitt's concept of the political and, many commentators believe, shapes contemporary policies and tactics of sovereign states and governmental agents. Schmitt's objective was to allow the moment of the ultimate decision to stand out. But at the origin of this sovereign decision lay a friend versus enemy opposition that was for Schmitt the quintessential political moment. While the decision on the exception and the manifestation of the friend/enemy opposition in concrete contexts were likely to lead to war, Schmitt still sought to argue that techniques and tactics of war were subordinated to this allegedly primordial antagonism. Thus, Schmitt insisted that his conception of the political was "neither belligerent nor militaristic, nor imperialistic, nor pacifistic."[59] At the same time, as we argued above, Schmitt's enemy could never be theorized separately from its relation to the constant and real possibility of war and war destruction. This was perhaps an inevitable outcome of Schmitt's theory since the enemy had to embody an existential threat to the *demos* and its way of life (it must be noted though that later on, in *The Nomos of the Earth* in particular, Schmitt switches to a concept of enmity that is more in line with the idea of *justis hostis*, or a just and honorable enemy, as we shall see in Chapter 3). Thus, as several scholars noted, the threat to the *demos* and its way of life coming from the concrete enemy could nonetheless be transmuted into a way of boundless death and destruction or, to be more precise, into a call to agonal warriors called in to perform glorious but potentially devastating deeds that do not need to respect the humanity of the *demos*.

Thus, it may not come as a surprise to read that German philosopher Karl Löwith emphasized the presence of an inescapable ambiguity at the heart of Schmitt's understanding of the relationship between politics and war. Indeed, Löwith asked: "Does the political exigency of war exist because there are essentially different peoples and states or political '*forms of* existence' which accord with a way of being; or is it only when war happens to take place … that even the most extremely tense and purely existential commitments and divisions emerge, which according to Schmitt are the distinctive and essential characteristic of the political?"[60] According to Löwith, Schmitt's thought actually was "amenable to *both* of these possible interpretations."[61] As a result, it would be impossible to decide within the Schmittian text whether war was fundamentally prior to enmity or whether, on the contrary, the antagonism between friend and enemy preceded the need to go to war. Thus, Löwith detected in Schmitt's concept of the political another dramatic paradox. This paradox turned into a full blown aporia when Schmitt's "decisive formulations of the enemy-friend distinction shift[ed] indecisively back and forth between a *substantial* and an *occasional* understanding of enmity … Upon the shifting ground of this ambiguity Schmitt builds up his concept of the political, whose essential feature is no longer the life in the *polis* but instead the *ius belli*.'[62]

If we push further Löwith's suggestion, the possibility that Schmitt's concept of the political—his notion of a sovereign decision to go to war to preserve a certain way of life or to defend a political unity—may always already be the least sovereign or the least quintessentially political moment there is, that is to say, an always possible and already conceivable total surrendering of sovereignty/political power to the agony of the war machine, leaves us with a reading of Schmitt's thought (and of its utility for contemporary critical theorizing) that is eminently undecidable or unresolvable. Nor is this reading resolvable in the context of Schmitt's later work, *The Nomos of the Earth*. For Schmitt's nostalgic reading of the rise and fall of the *Jus Publicum Europaeum* in *The Nomos* is predicated upon the physical demarcation of an outside—a particular space where individuals are deemed to inhabit a non-political/purely violent contingent domain that gives meaning and legitimacy to the bracketed wars between European sovereigns inside Europe—that is no longer properly recognized and respected in our contemporary international setting.[63] In fact, one may postulate that the instabilities or aporias encountered in Schmitt's earlier writings return in full force in his later works through his attempts at reconfiguring enmity in the form of borderless partisans.[64]

In concluding this chapter, and in opening the perspective on agonal sovereignty theorized here to the rest of the volume, we want to suggest that maintaining the undecidability or aporetic dimension of Schmitt's writings and their utility for contemporary reflections on biopolitics, necropolitics, or even agonal politics is of crucial importance. This is, we believe, what agonal sovereignty as an analytical perspective also allows us to achieve. Beyond academic debates over the merit of one critical analytical perspective (biopolitical or otherwise) over another, what matters is the capacity to shed light on the pervasiveness of war designs and discourses of war devastation in contemporary politics. What matters is to keep open and perhaps make public the debate on the role played by warring agents in social and political settings that have become instruments of violence, terror, and horror. Thinking critically along the lines introduced in this chapter enables us to peek into this abyss of violence, into the domain where war commanders and their heroic troops/tools operate without boundaries and beyond humanity as a limit condition.

For these reasons, the not-so-easily resolved tensions between war and sovereignty possibly found in Schmitt's texts (as Löwith suggests) need to be highlighted. Jacques Derrida has suggested that it is from the perspective of what he calls a "possible-eventuality" that Schmitt's work may best be appreciated. As it turns out, this "possible-eventuality"—or the plausibility of an occurrence that has already been realized and yet can never be fully assured in its realization—may be an important characteristic of the notion of war as it emerges from Schmitt's text. As Derrida puts it:

> As soon as war is possible, it is taking place, Schmitt seems to say; presently, in a society of combat, in a community presently at war, since it can present itself to itself, as such, only in reference to this possible war. Whether the war takes place, whether war is decided upon or declared, is a mere empirical alternative in the face of an essential reality: war is taking place; it has already begun before it begins, as soon as it is characterized as *eventual* (that is, announced as a non-excluded event in a sort of contingent future). And it is *eventual* as soon as it is *possible*.[65]

Perhaps, then, in its indetermination and fundamental instability, Schmitt's aporetic writing on the nexus war–politics–violence is a kind of prophesy. But it is a prophesy that is always open to contestation, never sure of itself, or indeed "possibly-eventual," as Derrida notes. Schmitt's prophesy here is also curiously reminiscent of Foucault's own efforts at evoking discourses of "politics as war." Indeed, like Foucault's writings, this Schmittian prophesy reminds us that war is always already taking place in any politically normative or regulative context (decisionistic or deliberative, authoritarian or plural).

It recalls for us the inescapable fact that war and its agonal agents have begun their long destructive march towards death, terror, and horror from the very moment when some political process, entity, actor, or even idea has characterized them and their glory as "possibly eventual." It is this possible eventuality of war, of the war machine, and of its dehumanizing violence that, among other things, agonal sovereignty seeks to explore. We do so, in the next chapter, by considering how fear and its connection to power and force operate in a context of horror.

2 Nothing to fear but fear itself

Governmentality and the reproduction of terror

Introduction

Whereas late-modern societies thought they had conjured away the specter of a state-of-nature-like violence, events of the past decade have reminded people in developed, rich, and supposedly danger-free Western nations that physical harm and material destruction can hit anyone, anywhere, at any moment. Since the attacks of 9/11 (New York and Washington, D.C.), 3/11 (Madrid), and 7/7 (London), people in the West have come to the realization that, in the words of Zygmunt Bauman, they are living "the terrifying experience of heteronomous, vulnerable populations overwhelmed by forces they neither control nor truly understand, horrified by their undefendability and obsessed with the security of their borders."[1] An important part of this apparent accentuation of fear "at home" (and not just "over there") has been the perception that today's enemy, envisioned as an amalgamation of forces often referred to as terrorism and potentially present everywhere,[2] is an enemy like no other in history (we will return to the question of enmity in Chapter 4). The acts of 9/11, 3/11, and 7/7, among others, have popularized the idea that the enemy is beyond any form of restraint, located in a domain where it may no longer obey any form of political rationality. In 2005, George W. Bush did his part to anchor this type of danger by claiming that the terrorizing enemy is "as brutal an enemy as we've ever faced. They're [sic] unconstrained by any notion of our common humanity, or by the rules of warfare."[3]

If one of the main arguments for the justification of the modern state, since Thomas Hobbes' writings, has been the ability of the sovereign state to defend the property and the basic rights of individuals (starting with the right to be free from harm), different questions about sovereignty—questions possibly rendered more urgent by events of the past decade—must be asked. For example, what happens to practices of sovereignty when the fear of the enemy borders on the absolute? What do sovereign power and force become when fear/terror targets life itself? Perhaps more crucially, what supplementary configurations of violence accompany the gruesome acts of warring heroes when war and the war machine develop their own "sovereign" logic, as we saw in Chapter 1?

The Hobbesian recipe to achieve order and security, which consisted in creating a sovereign power with a primary capacity to offer protection from violence through a concentration or centralization of fear, is well known. For Hobbes, the solution to chaos was to place the source of fear in the very persona of the sovereign. Yet, as Hobbes' state of nature was supposed to serve as a constant reminder of, when fear escapes the sovereign domain and becomes disseminated throughout the social compact, absolute terror and the specter of total annihilation easily could return. Such a fear of catastrophic terror was not only a fear of potential physical harm or possibly violent death. It was also crucial to the maintenance or even reinvention of modern political institutions (still centered around the idea of state sovereignty) as fear allowed those who mobilized its specter periodically to call into question practices that nonetheless accompanied the development of the modern state to better (and more rationally) reconstruct them. Thus, the discourse of modern sovereign power and force has traditionally been about finding new ways to re-anchor the authority of the state and its institutions, often by calling in fear as a trigger for change. For example, over the past two decades, the so-called eliding of traditional geopolitical boundaries, the multiplication of deterritorializing flows associated with the presence of stateless groups or entities (bands of warriors, terror networks, cartels, wandering refugees, and so on) no longer under the control of state institutions, or the observation by some of a growing global disorder as a result of both natural and man-made catastrophes possibly leading to generalized civil war conditions may have contributed to the notion prevalent among Western state leaders (and sometimes their political subjects too) that a transfer of sovereign prerogative from the centralized state to its order-enforcing or risk-managing agents (police forces, the military, immigration officers, and so on) is needed. Although these ideas and perceptions are not new, the resurgence of discourses and images of total terror in the wake of 9/11 has led political theorists to notice the propagation of all sorts of executive and decisionistic measures and policies that often appear to stand in for the sovereign decision.[4] Still, many of these measures and policies have not been the result of a sovereign decision (as Carl Schmitt would argue), but rather of techniques and tactics deployed by various public agents/agencies in charge of recapturing security, order, and the law in the absence of a central sovereign power that could concentrate fear in one place (or, at least, with such a power now taking a secondary role in contemporary political decisions).[5]

Hobbes' famous solution to what he perceived to be the danger of civil disorder and chaos once again was to create a power that could centralize all individual fears but also all the methods devised by individuals to react to such fears. The arbitrary, executive, and policing force of the one and undisputed sovereign (thanks to whom all individual subjects could finally achieve security and well-being) would be the sole guarantee that the "natural" danger of violent death could be done away with, even if that meant that organized violence against other sovereigns (and their populations) had to be undertaken to preserve the domestic compact. Hobbes thus became a paradigmatic figure for subsequent theories and models of centralization of sovereign power, a power nonetheless based upon the idea of an always possible overwhelming fear of danger, chaos, and destruction.

In the twentieth century, Hobbes' centralized state power would be revisited by Schmitt's own founding of the state on the sovereign's prerogative to mobilize a series of conditions and situations of exception. For Schmitt, the legal and constitutional order of the modern state had to have the capacity to cast itself out through an act or decision of the executive, but also, potentially, through the actions of policing or soldiering agents (as we saw in Chapter 1), during a period necessary (though supposedly short-lived) to defeat a deadly enemy. At first glance, Schmitt's sovereign decision on the exception can be read as the logical outcome of Hobbes' desire to preserve the domestic compact at all costs, and to allow society to thrive through mechanisms that can ensure control over fear/danger.

The first part of this chapter briefly reviews how Hobbes and Schmitt organized their thoughts with a view towards perpetuating techniques of centralization of fear around the institution of the sovereign in modern politics. Still, today, and as suggested above, the Hobbesian mechanism of fear concentration and the Schmittian condition of sovereign exception often look like they have been generalized in global politics. Instances of war against and destruction of enemies proliferate across the globe in the hope that such antagonistic and supposedly exceptional operations will make "us" free from fear "in here."[6] In this context, the Global War on Terror has been a convenient excuse for Western states, since it has authorized them to execute openly the kind of boundless violence in the name of an eradication of danger that, for centuries, they practiced with proficiency against the non-West. In the West, a typical response to the threat of so-called dangerous enemies has consisted in re-injecting in the sovereign state a capacity to institutionalize fear at any moment by having recourse to war and warriors. As a result, as we noted in the previous chapter, an agonal sovereignty for which the preservation of the political compact is no longer the final objective has taken over.

However, in this chapter, we also seek to move past Hobbes' and Schmitt's solutions to the "fear production" dilemma by again introducing some of Michel Foucault's analyses, particularly those on governmentality.[7] Foucault allows us to identify a supplementary but crucial dimension in the relationship between fear and modern power. To be sure, Foucault's concept of power is different from that introduced by Hobbes and expanded upon by Schmitt. As is now well-known, Foucault stresses a non-essentialist and relational concept of power that further allows him to map out a non-sovereign centered potential for disciplinization and normalization. At the same time, a less debated aspect of Foucault's concept of power is its connection to practices and discourses of war, as we noted in Chapter 1. Foucault's micro-histories of Western discourses of war, battles, and racialized antagonistic violence suggest that fear (and the power relations that flow from its production) is actually not something that the modern state and its agents ever want to do away with or be free from. Rather, fear is what must be produced and reproduced by governmental agents in order to establish the control, supervision, or enhancement of the social body through multiple mechanisms of measurement, calculation, improvement, and preservation of life. Thus, Foucault intimates, fear must be made productive and reproductive in and of society, not only to allow the sovereign state to mobilize death, terror, or endless destruction through a recourse to war and warriors, but also, apparently, to enable life—or a certain conception of what it means to have live bodies in society—to thrive.

It is indeed through a series of governmentalized techniques or procedures of maintenance of life that, in the modern age, fear has been made "beneficial" to society by, first, operating at the level of individual docile bodies (through disciplinary mechanisms) and later, around the turn of the nineteenth century, working on the population in its entirety (through methods of rational regulation). Thus, one could argue that by the time Schmitt (in the twentieth century) seeks to revisit Hobbes' model of sovereignty (as a system of power premised upon a concentration of fear), a generalized biopolitics of fear has already been put to good, efficient, and "positive" uses in the modern state, through the disciplining and normalizing efforts of various governmental agents that may or may not directly serve the interests of the central sovereign/executive power. Such a perspective is clearly in line with biopolitical analyses, and it intimates the presence of what some have started to refer to as biopolitical sovereignty (as we saw in Chapter 1).

As agencies, arrangements, or assemblages of surveillance and regulation of bodies and the population disseminate their effects throughout the body politic, governmentality displaces authority and power away from the centralized sovereign.[8] More than a politics of sovereign exceptionality, it is a *biopolitics of fear* enacted by way of governmentality that is operative and that, in a way, disables the state's central monopoly on power. This pluralization of fear and power in governmentalized modernity further encourages all sorts of public agents/agencies to mobilize the specter of danger, threat, insecurity, and enmity. Far from mastering the conditions of production and reproduction of fear (as Hobbes, Schmitt, or even some contemporary proponents of a return to sovereign exceptionalism would have it), the sovereign is actually made to depend upon a wide array of decentralized "executive," sometimes public, and generally administrative procedures and mechanisms (or *dispositifs*, as Foucault would call them) that bear the mantle of social order and security. A Foucault-inspired study of the way fear is rendered productive in modernity (or a biopolitics of fear) seeks to create a rupture in analyses that have suggested that Schmittian theories of sovereignty are best suited to explain the return to a politics of fear today. Indeed, informed by a biopolitical frame of representability, such an analytical approach suggests the presence of a political and discursive context in which multiple governmentalized agencies proliferate power-effects, control-effects, security-effects and, ultimately, terror-effects throughout society.

According to a biopolitical approach to fear and terror, when biopolitical agents/agencies of fear production—police forces, the military, immigration and customs officers, airport security services, but also some educators, some doctors and scientists, some legal and constitutional experts, or some administrators of public bureaucracies—become the loci of enunciation of techniques of governmentalized power, the likelihood of unlimited violence and the prospect of a generalized condition of terror are no longer what must be cast away. Rather, a shift is taking place whereby these conditions are what must be expected, accepted, and anticipated by populations whose lives are said to be constantly threatened and, as such, must become the objects of sustained normalization or heightened regulation.

Still, in the last section of this chapter, we push further this idea of a reversal in the productive mobilization of fear. We examine the thought process and the logic that enable proponents of a biopolitics of fear/terror approach to theorize the passage from a fear of violent death to a fear of letting the "wrong" people live and, consequently, of not being able to allow the safe and regulated populations to enjoy "normal" lives. In particular, we revisit Mick Dillon's notion of "emergent life" and we argue that it is closely tied to the perspective on agonal sovereignty and horror (not only terror) we described in Chapter 1.

Hobbes, Schmitt, and the centralization of fear

In the Hobbesian model of power, fear is presented as an inescapable condition of physical violence and of moral/mental angst.[9] It imposes a singular obligation on the sovereign: to protect the members of the social compact. To some, this obligation to protect rights-bearing individual subjects positions Hobbes as the first truly modern political philosopher.[10] From this perspective, the Hobbesian state of nature is a crucial starting point for any modern political thought that seeks to empower a distinctly "artificial" unity, or what for Hobbes becomes the modern sovereign. In *Human Nature*, Hobbes offers an elaborate and systematic overview of what human nature is. Hobbes writes: "Man's nature is the sum of his natural faculties and powers, as the faculties of nutrition, motion, generation, sense, reason &c. For these powers we do unanimously call natural, and are contained in the definition of man, under these words, animal and rational."[11] Hobbes seeks to show that the first impulse towards human action in nature is rooted in a "conception of external objects," one that "causes appetite" and by extension "fear."[12] Thus, when Hobbes examines the implication of a will rooted in desire, appetite, and the concept of fear, what emerges is a fundamental "*right of nature*: that every man may preserve his own life and limbs, with all the power he hath."[13]

Hobbes attempts to rationalize society on the basis of a new capacity to mitigate the worst form of fear that inevitably (he believes) occurs between men and of a likelihood of great physical harm resulting from human vainglory, competition, and diffidence. The state of nature as a state of war is one where "there is in all men a will to do harm which derives from vainglory [*inanis gloria*] and over-valuation of his own strength."[14] This state of nature/war, famously characterized by Hobbes as "solitary, poor, nasty, brutish, and short,"[15] becomes the necessary platform for the legitimization of the overarching power of the "mortal god," or Leviathan, that appropriates the mutual fears among individuals and concentrates them to produce order and security. Hobbes further justifies this concentration of power/fear by arguing that the defense of the political realm against hostile outsiders necessitates "*one Assembly* or *one man* who has the right to arm, muster and unite, on each occasion of danger or opportunity, as many citizens as the common defence shall require … as well as the right to make peace with the enemy when advantageous."[16] Of note here is the fact that fear—the fear of violent death in particular—always remains the currency of the political legitimacy and constituted rule of the sovereign, even after the social compact is created and allegedly stabilized. When Hobbes writes that "the nature of war consisteth not in actual fighting, *but in the known disposition thereto*,"[17] he articulates the belief that fear must remain the propelling force of any political system, perhaps at any stage of this system's life. But the fear Hobbes capitalizes on to legitimize Leviathan must also remain rooted in the virtuality of future threats against the polity. What can be called a virtual condition of fear (fear can return at any time because disorder is virtually present) plays a key structuring role in the Hobbesian state (we shall return to the issue of virtuality in Chapter 3). Hobbes emphasizes "a tract of time, wherein the will to contend by Battell is sufficiently known: and therefore the notion of *Time*, is to be considered in the nature of Warre … "[18] What gives rise to the "*known disposition thereto*" reflects a concern not only with actual possibilities of violence, but also with the intent of those who may pose a threat. Thus, Jef Huysmans has suggested that Hobbes relies upon an "epistemological fear." Accordingly, "insecurity does not follow from one's vulnerability as such. Rather, it follows from an uncertainty about which human relations are benign and which are dangerous."[19] As a result, the centralization of fear in the institution of the state is a necessary step for Hobbes because it serves as a prophylactic device against the overwhelming fear of virtually present threats.

Despite the apparently durable constitution of this centralized system of power, the status ascribed by Hobbes to the state of war remains highly uncertain. Does Hobbes imagine the state of nature/war (and the threat of its recurrence, even after the institution of Leviathan) as a real possibility, as an ongoing danger to political life? Or is it mainly, as some have argued, a heuristic device,[20] contaminated by assumptions about human nature developed within an already present society, and thus designed to justify the sovereign's authority and its monopoly on violence? In other words, is the "terror" of the state of nature a concrete condition, one that any society can encounter or fall back into? Or are we instead to interpret Hobbes' political philosophy as a prime example of the type of fear production often mobilized by states and their leaders to maintain or reinforce sovereignty (a sort of always already anticipated state of exception)? These questions become more pertinent when we summon Schmitt's political theory and Schmitt's desire to accentuate the terror supposedly found in the state of nature. For without a persistent sense of fear present in the minds and bodies of individual subjects, the awe-inspiring power of the state/sovereign could lose much of its legitimacy, and the ultimate ability to make crucial executive decisions to safeguard the political compact may no longer appear justified.

By the time we get to Schmitt's revival of Hobbes' thought on the centralization of fear in the hands of the sovereign, the emphasis has been shifted to sovereignty as the pinnacle of political order. Schmitt has been described as a paradigmatic figure in the twentieth-century revival and perhaps radicalization of Hobbes' model of power.[21] In part because of his perception of the political chaos that reigned during Germany's Weimar Republic in the 1920s, but also due to what he considers to be the negative effects of liberalism (with, among other problems, liberalism's neutralization of political life),[22] Schmitt seeks to revive the terrifying specter of a return to the state of nature. He presents this fatal reversion as a situation that could only result in the demise of the state.[23] Schmitt's two main works at the time, *Political Theology* and *The Concept of the Political*, attempted (among other things) to reset the terms of the debate over the role of fear in power configurations by postulating two important principles: (1) sovereignty is the capacity to determine when an exceptional situation arises;[24] (2) the basis of the political lies in the ability of the sovereign to make a decision as to who the public friend and the public enemy are.[25] For Schmitt, any inquiry into the state must first reckon with the concept of the political and these two principles.[26] We have already discussed these principles in Chapter 1. Moreover, like Hobbes, Schmitt insists on the fact that "to the state as an essentially political entity belongs the *jus belli*, i.e., the real possibility of deciding in a concrete situation upon the enemy and the ability to fight him with the power emanating from the entity."[27] However, only when the sovereign state retains a central capacity to marshal authority, resources, and legitimacy does it have the ability to make such an essential decision. And, crucially, only when the population retains a capacity to fear the consequences of a return to the state of war can the state truly be political.

Schmitt is interested in making use of Hobbes' fear (and his state of nature/war specter) to normalize a mode of sovereign decisionism, a foundational decision on the exception that, in fact, necessitates a constant production of fear of physical harm or death. Without such an unremitting anxiety, the rationale for abstracting any normative content out of the juridico-political order would disappear, and what Schmitt perceives as a form of partisanship (with respect to the determination of the legal substance of the political order) would take over the state and deprive it of its vitality. This is perhaps why there is always a need for Schmitt to think in terms of extreme cases, or to mobilize the figure of an absolute threat (even though he claims to refuse to think of the enemy as absolute), or to imagine what is in excess of the political, as Derrida suggested.[28] Such a way of thinking also seems to justify the deployment of political and juridical concepts that revolve around the idea of a decisionistic unity. What Schmitt dreads most is the impotence of the state. Whether Schmitt realizes there is always a dramatic possibility that the sovereign could end up exchanging the terror of the state of nature for a terror brought on by war-mongering agents remains debatable (see our analysis in Chapter 1). In fact, his decisionistic model is what is likely to lead to a state of affairs where a monstrous executive power (or agent yielding arbitrary but awesome power) and representing an unbounded will takes over political life. This is the kind of scenario that, as Žižek argued, would normalize "abyssal violence" (as once again we indicated in the previous chapter),[29] that is to say, a political action that is no longer restrained by any form of obligation to secure the individual subject or the entire political compact, and whose consequences for the state itself would no longer be controllable.[30] However, Schmitt's objective is to neutralize anything that could compete with the state's monopoly not simply on violence, but also on the production of fear and terror. In this fashion, and in an agonally sovereign manner, Schmitt ends up conflating the sovereign decision to go to war and the always possible return to the state of war in the image of a Leviathan-turned-Monster of War for whom chaos or destructive violence and order/preservation of the state can become one and the same.

As some commentators have noted, the image of this sovereign monster has had a tendency to return in recent political circumstances.[31] A normalization of terror, violence, and fearful life is often the outcome of many of this sovereign's actions and decisions (or, better yet, of the actions and decisions taken by the sovereign's heroic warring agents). Contemporary critical studies have been tempted by this analysis, particularly in the context of the US-led war on terror.[32] By contrast, another critical approach influenced by some of Foucault's writings suggests that a careful understanding of the way mechanisms of normalization operate requires one to de-emphasize the central role given to the sovereign in political designs. Indeed, without a removal of the emphasis on the central role of the sovereign in the contemporary politics of force, fear, and violence, critical perspectives would remain wedded to an essentially Hobbesian (if not Schmittian) model of power, one that automatically assumes a hierarchy of authority and the institution of a fixed mechanism of violence concentration. It is towards Foucaultian considerations and their implications for a governmentalized and possibly biopolitical approach to fear production that we now turn.

Foucault and the biopolitical production of fear

Despite Schmitt's alleged radicalization of Hobbes' thought, Hobbes and Schmitt do seem to share a determinant point of view when it comes to power and sovereignty. For Hobbes and Schmitt, sovereign power is rooted in the constant possibility that the state may decide to take away human life. This ability to take away life or to put to death is presented as a functional mechanism of sovereign power that serves to establish or reinforce the clear distinction between the state on the one hand and society or the citizenry on the other. This juridico-political framework is constructed as a continuous institutional system of monopolization of force that enables the singularity of the sovereign decision to be readily operationalized and, in some cases, turned into an awe-inspiring deployment of centralized violence and terror (even when the sovereign claims to use force to protect the rights of the citizens). This always present possibility of violence is what ultimately guarantees the legitimacy of the modern sovereign or state. But this powerful configuration also makes fear the necessary currency of the modern juridico-political order.

Within this modality of power lies another potentiality, one that cannot be reduced to a juridical rule or to the preservation of the legal order (even by way of a decision on the exception). Foucault's early work on disciplinary power points to the way disciplines invest political, social, and economic institutions and procedures beyond the immediate gaze of the sovereign.[33] The disciplines' micro-powers and their meticulous control and management of bodies, tied to a series of new knowledges (medical, penal, pedagogical, and so on), appear to create a rupture with the Hobbesian/Schmittian attempt at placing the will and force of the sovereign (and, often, the fear of such a will or force) at the heart of modern mechanisms of political order. Foucault distances himself from the Hobbesian view on power and offers instead a relational perspective. He argues that "[p]ower must … be analyzed as something that circulates … [since it] is never localized here or there." Foucault adds: "Power passes through individuals. It is not applied to them."[34] Thus, whereas Hobbes stresses the virtuality or, at least, the future-oriented potential for chaos and all-out war as a key factor in the legitimization of the state, Foucault's genealogies of power navigate through actual historical wars and war accounts that gave rise to (as we saw in Chapter 1) "historico-political discourses" of racial struggles.[35] This contrast with Hobbes found in Foucault's genealogies of sovereign power is important because of Foucault's emphasis on revealing the "historical contents that have been buried or masked in functional coherences or formal systematizations."[36] In other words, as Foucault clarifies at the beginning of *"Society Must Be Defended"*, genealogies are meant to be "an insurrection against the centralizing power-effects that are bound up with the institutionalization and workings of any scientific discourse."[37] As a result, Foucault brings to light what Hobbes' model of centralized power and fear sought to suppress: the real-life battles and political struggles that gave rise to the law, states and nations, and the vast array of modern apparatuses of government. For Hobbes, the emergence of Leviathan is tied to "the political use that was being made in political struggles of a certain historical knowledge pertaining to wars, invasions, pillage, dispossessions, confiscations, robbery, exaction, and the effects of all these feats of battle."[38]

When Foucault addresses sovereignty and the formation of modern state apparatuses, he seeks to introduce the notion of biopower, in part to mark a break from the Hobbesian conception of sovereignty, as suggested above.[39] With Foucault's concept of biopower (described in the "Introduction" to this volume), the point of reference is no longer the forceful, centralized, or awe-inspiring rule of the sovereign. But it is also not just the patchwork of sites and regimes (the asylum, the hospital, the prison, the school, the factory, and so on) that seem to characterize disciplinary power, as Foucault first introduced the concept.[40] Rather, and to repeat some points stated earlier in this book, the object of biopower is the population of a given territory in its entirety, and the main preoccupation of this power is the population's proper governance. This shift from a centralized sovereign model of hierarchical power to one that conceptualizes relational techniques of governmentalized biopower has been described by Foucault as a transition from a power that defined itself by way of "the right to *take* life or *let* live" to one that seeks "to *foster* life or *disallow* it to the point of death."[41] The emergence of biopower as a primary governing logic (around the eighteenth and early nineteenth centuries) is also intended to give rise to new forms of knowledge about and around the social body. Rather than being overtly concerned with contingent events such as reacting to an epidemic, targeting particular "natural" abnormalities, or fending off local disorders, biopower focuses on "what might be broadly called endemics, or in other words, the form, nature, extension, duration, and intensity of the illness of the population."[42] These "endemics" are more or less durable features of a population, a species, or even a race. They are said to be responsible for a series of social factors that can impact negatively the vitality or health of an entire population, society, or body politic.

This passage from a fear of immediate death at the moment of the always possible catastrophic event or of the sovereign's reaction to it, to a concern with maximizing conditions that can foster life in society reveals a transformation in the political utility of death in modern power designs, something that once again we mentioned in the "Introduction." With biopower, death becomes "something permanent, something that slips into life, perpetually gnaws at it, diminishes it and weakens it."[43] While the termination of individual lives is an inevitable feature of natural life, in both biological and political discourses death now becomes a matter of mortality, something that counts at the level of the entire population.[44] With the notion of mortality (and its social, economic, political, but also biological connections to the overall vitality of the social body) regulatory norms emerge. Their purpose is to ensure that a social equilibrium can be maintained or can compensate for fluctuations that are still bound to arise from time to time.

What Foucault calls "security mechanisms" can be understood in the context of this biopolitical organization, normalization, and maximization of the conditions of proper governance of life in society.[45] The implementation of security mechanisms or safety procedures throughout the social (and the multiplication of agents and agencies in charge of operating them) revolves around the idea of not only investing individual bodies (as the disciplines did), but also of optimizing the health or labor functions of a population.[46] In this sense, the rise of biopower for Foucault also points to the growing difficulty for the centralized sovereign to master the novel political configurations, economic situations, and socio-physiological contingencies of life that emerge in the wake of the Industrial Revolution and as a result of the population growths of the eighteenth and nineteenth centuries in Western Europe (or what regimes of biopower will redefine as matters of "demographics").[47]

In the last chapter of *"Society Must Be Defended"*, Foucault tries to reflect on some implications of this turn towards biopolitics and security mechanisms. One important implication is that the function of state power appears to shift from one that is in charge of instituting a legal order over a particular political territory, to one that consists in delineating a succession of governmental procedures, techniques, and strategies that can actively regulate a population. But if state power as centralized violent force or coercion is indeed re-imagined, redesigned, or redistributed throughout biopolitical arrangements, power's relation to fear may have to be revisited too. If and, if so, how such a power over life relates to the modern production of fear, and whether, by seeking to optimize the living conditions of a population, biopolitics can or even is meant to remove fear once and for all now emerge as crucial questions. Or, to present the dilemma of biopower somewhat differently, how a biopolitics of fear may still be needed for the production of a new modality of power becomes a key problematic.

Although never directly posed by Foucault, these questions are nevertheless prompted by his reflections. If fear plays a part in biopolitical designs (something Foucault implies), it is no longer to prevent a transgression of the juridico-political order maintained by a central sovereign. Rather, it would have to be to preserve or enhance the life-efficiency of a given population. Thus, a new productivity of fear/terror seems to accompany the deployment of regimes of biopolitical governance (in the eighteenth and nineteenth centuries in particular, and possibly up until today). This productivity of fear and its connection to biopolitics make sense, Foucault suggests, if we take into account the way race comes into play in practices and discourses of power in the seventeenth and eighteenth centuries. Foucault argues that, around that time, Western European discourses of racial antagonism—found not just in biological studies but also in war declarations, political treatises, or nationalist pamphlets—start to introduce a knowledge that seeks to provide criteria regarding who can or should live and who, on the contrary, can or should die. This biopolitical discourse of race and racial enmity (a precursor to the idea of representing the enemy as a racial other) hopes to dismantle any understanding of the human species as a biological continuum, and it strives to impose hierarchical divisions based on categorical racial differences. Geopolitically, this discourse and subsequent practice of biopolitical racism attempt to re-territorialize the globe along the lines of such racial profiles and differences, a process that helps to "naturalize" the Western colonization projects of the eighteenth and nineteenth centuries. Foucault concludes that this biopolitical racism changes the "relationship between my life and the death of the other [in] that it is not a military or warlike relationship of confrontation, but a biological-type relationship."[48]

The normalized but supposedly "secure" society ("secure" from the racial other, and with normal living conditions that are not those of the other) that is created by what Mitchell Dean has called an "enwrapping of the modern state" by biopower relies a great deal on representations and operations of biological racism.[49] In many ways, this generalized yet intricate racial discourse of normalization (of the inside) and of antagonism (of the outside) does not just work through a differentiation between races, but also through all sorts of marginal forms of conduct, such as criminality, madness, underclass habits, and so on.[50] Thus, the normalization of the inside (the population, society) and the antagonism of the outside (the racial other, the enemy) increasingly become blurred. The ways the state and its agents start to conceive of relations with "other races" or with various forms of "counter-conducts" become similar and undistinguishable. In all cases, it is a relation of permanent hostility (or, perhaps, "endemic enmity") that is being implemented and that demands a productive disposition on the part of the population vis-à-vis antagonism, enemy-construction, and moral and physical confrontation.[51] As Foucault puts it: "[f]rom this point onward, war ... is not simply a matter of destroying a political adversary, but of destroying the enemy race, of destroying that [sort] of biological threat that those people over there represent to our race."[52]

The nexus biopolitics-race-war highlighted by Foucault allows us to revisit the relationship between power and fear in modernity. Instead of asking how the sovereign can manage to keep dangers and fears in check, a more pressing question seems to be to identify the extent to which biopolitical techniques and mechanisms of government are indicative of a productive fear of not being able to live one's normal, regulated, or optimized life as a member of the public, population, or society. What the perspective on biopower and the desire to preserve or even regenerate one's own race also suggest is the possibility that various decentralized agents/agencies of government will need to make efficient use of fear (once again, a fear of not being able to live one's normal life as part of a given population). These governmentalized agents and agencies can still have recourse to violent techniques or brutal force to protect a certain society. But, more often than not, they will only need to rely on "benign" calculated, rationalized, and calibrated safety measures or security mechanisms that supposedly leave no physical marks on individual bodies since they operate at the more abstract level of the organization, normalization, and optimization of the body politic.

With the emergence of those agents/agencies of government and their constant efforts to regenerate the population through a generalized sense of fear of not being able to live one's own life in an era when racial others or unproductive counter-conducts abound, we enter the realm of what Foucault refers to as governmentality. For Foucault, governmentality represents the moment in configurations of power when methods of scientific, disciplinary, and knowledge-based rationality and procedures of government come together to organize social life, often "through a range of formally nonpolitical knowledges and institutions."[53] The "governmentalization of the state" (as Wendy Brown puts it[54]) implies that this government, by institutions and knowledges, can pluralize the supposedly centralized sites of sovereignty and state power. With governmentality, the origin of the law is less relevant to social and political designs (and to power's applications) since this particular modality of government is "not a matter of imposing laws on men, but rather of disposing things, that is to say, to employ tactics rather than laws, and if need be to use laws themselves as tactics."[55] It is indeed through techniques and tactics of management and by way of organizational and procedural *dispositifs* that governmentality takes charge of a population, fosters its vital conditions (by trying to control endemic social "illnesses"), and "orchestrates the conduct of the body individual, the body social, and the body politic."[56] The "conduct of conducts" within the realm of biopolitics is the main objective of governmentality's social, economic, and political, but also cultural, religious, and educational agents and agencies. In a context of governmentalized power, the biopolitical production of fear is the result of a series of scare tactics or terror *dispositifs* put to "good" social effects by agents/agencies of government.

The Foucaultian concept of *dispositif* (often translated as "apparatus" in English) is particularly useful to an analysis of biopolitics and fear. *Dispositifs*, Nikolas Rose explains, are "machines for government."[57] They are complex apparatuses made up of various economic components, political unities, discursive elements, and bits of socio-cultural information that come together for strategic reasons,[58] not to create normative structures or to maintain legal orders, but to produce power as governance effects. As Giorgio Agamben has noted, *dispositifs* demonstrate a "capacity to capture, orient, determine, intercept, model, control, or secure the gestures, behaviors, opinions, or discourses of living beings," often with a view towards achieving some level of subjectification.[59] Moreover, the strategic or tactical work of *dispositifs* is more likely to be justified (by agents of government) when public discourses about "urgent needs" proliferate since *dispositifs* or apparatuses can "obtain an effect that is more or less immediate."[60] Thus, the power of *dispositifs* has to do with the operational and regulating strengths they inject into governmentality. *Dispositifs* can also be thought of (as Gilles Deleuze suggests) as "lines of forces" that seek to relate various points in society without any necessary pre-existing commonality, connectivity, or associative logic.[61] To the extent that it makes sense to use the phrase "agents or agencies of governmentality" (as we do in this chapter), those agents/agencies and their tactics of regulation, normalization, and optimization of life can be seen as layers, collections, or assemblages of various *dispositifs*. *Dispositifs* are indeed machines *for* government (but not *of* government) whose regularities of conduct nonetheless owe very little to the possible existence of a centralized power or to the traditional model of sovereignty.[62]

If fear is one of these regularities generated by *dispositifs* of or for government (if fear becomes a terror *dispositif*, in other words), fear cannot be the outcome anymore of what the sovereign wishes to contain, control, or localize in some place (in the state of nature, in the executive decision of the sovereign, in the persona of the king, in the constitution). Rather, a biopolitical production of fear implies that a multitude of agents *for* government are in the business of deploying a range of techniques, procedures, and tactics that strive to realize control over or regulation of the population's everyday conduct from all sorts of points, positions, or perspectives and through various relations or lines of force. It is everywhere throughout society that we find security mechanisms or safety arrangements that are constantly at work mobilizing fear and terror. In the next section of this chapter, we reflect on some of the meanings of this biopolitical analytical approach to the governmentality of fear in contemporary settings of social terror. More importantly, we also reveal the point at which this biopolitical perspective stops to make sense of the current condition of terror or, perhaps, even horror.

Scare tactics, terror dispositifs, and the emergence of a life primed for horror

In a recent essay, Didier Bigo also turns to Foucault's notion of *dispositif* to situate post-9/11 security practices within the general field of governmentalized power. For populations, the production of (in)security generally amounts to making use of situations, events, or contingencies that can result in individual bodies being made incapable of or in fact prevented from enjoying their normal life or expected living conditions. What Bigo calls the "management of unease" is one form the biopolitical production of fear takes in contemporary society (after 9/11, above all), both within and across nation-state borders.[63] This production of unease or, to put it differently, the governmentalization of insecurity in contemporary society, highlights the omnipresence of decentralized political operations through terror *dispositifs* or, perhaps, scare tactics.

To make sense of contemporary *dispositifs* of fear/unease within a generalized condition of governmentalized insecurity, Bigo revisits Jeremy Bentham's famous panoptic regime of surveillance (discussed by Foucault in *Discipline and Punish*),[64] and he comes up with a new concept: the "Banopticon." This "Banopticon" operates not just by means of disciplinary omnivoyance over coerced bodies, but also through normalized exclusion (or *ban*) of targeted individuals and groups. Some of these targets are marked as racially other or as representing exemplary counter-conducts and, as such, become dangerous enemies for the population. However, many banned individuals and groups are not characterized *a priori* as abnormal, criminal, delinquent, inimical, or even terrorist. Under a "banoptic" regime, anybody can become the object of banning micro-gazes or micro-surveillance tactics mobilized by all sorts of apparatuses involved in the population's security or well-being. For Bigo, the "Banopticon" intervenes at the juncture of governmental rationality and power's capacity to remove the rule of law in times of emergency. This *dispositif* of terror and management of unease is also supposed to anticipate the "future potential behavior" of individuals and groups and to normalize "the non-excluded through [a] production of normative imperatives."[65]

Bigo's analyses of unease are intriguing even though they tend to privilege a unidirectional perspective on governmentality and fear. Indeed, fear here appears to be managed at the governmental level (through terror *dispositifs*), and it then seeps into the public sphere through a relentless process of normalization. But, unwittingly perhaps, a vertical structure of power is reproduced by this understanding of "banoptic" systems of governance and exception. As suggested above, any "banoptic" exercise implies that a condition of exception or a state of emergency has been deployed. More crucially, what is absent from Bigo's creative reflections on governmentality and unease is an elaboration of how the live but banned bodies of the population, under "banoptic" conditions, internalize or, rather, are encouraged to make their own a fear of not being able to live according to normal or expected conditions. A critical concern about contemporary regimes of governmentality is not just why individuals and groups in society fear not being able to live their normal life. It is also how the myriad apparatuses through which rationalities of government are rendered visible and effective manage to make fear/terror the *population's own problem*, that is to say, how they produce fear/terror as an everyday social reality that populations must own and become responsible for.

Asking how the productivity of fear by way of agents/agencies of governmentalized terror is realized means questioning the ways in which a biopolitics of fear, unease, or insecurity takes place *and* reproduces itself across space and over time. It means wondering how the governmentality of fear develops today the conditions of production of its eventual reproduction. And at this point, we can start to perceive another limitation of biopolitical frames of analysis. Indeed, whether we think about this management of unease or fear—and the methods through which it is kept active—in terms of special sites or exceptional events (Guantanamo's Camp Delta after 9/11), or as a matter of everyday occurrence in public places or through mundane activities (commuting to work via the London Underground transit system in the aftermath of the July 7, 2005 attacks),[66] exploring how human bodies are mobilized to provide their own responses to the governmental management of fear/terror in the particular timescapes of (bio)power is crucial to assessing the resilience of contemporary regimes of power, force, violence, and horror. Human bodies primed for an acceptance of trauma and ready to take upon themselves measures and technologies that can maim, disfigure, or bludgeon their own flesh (and, as such, are also prepared to negate their humanity) are the kind of organisms and cognitive response systems regimes of fear and terror under horrific conditions operate with and through these days.

Dean writes that "[o]ur present is one in which we are enjoined to take care and responsibility *for our own lives, health, happiness, sexuality, and financial security.*"[67] As Dean intimates, the governmentality of fear, unease, terror, or insecurity demands that subjects/bodies act upon themselves first and foremost. To borrow Dean's language, it is one's own "enfolding" of and in governmental authority, the application of *dispositifs* of fear and security by one's flesh and into one's flesh, that becomes the best guarantee that regimes of governmentalized terror will remain effective. This embodiment or, better yet, incarnation of "banoptic" techniques is a preferred disposition of contemporary fear production.[68] It is by means of such a localized, disseminated, disaggregated, and perhaps dismembered operation that the quotidian politics of terror is best maintained and reproduced.

Put differently, if insecurity or unease is the problem, and if fear/terror *dispositifs* are to be the answer, governmental agents/agencies must certainly make sure that the conduct responsible for the alleged irregularity or abnormality is done away with (before it turns into an endemic, as we saw above). This is once again what biopolitical and governmentalization centered analyses show us. But, more importantly, these terror *dispositifs* upon which governmental agents rely turn to bodies or organisms that are actively recruited in the maintenance of these regimes of governance. It is precisely these bodies, organisms, or amalgamations of human flesh that must self-rationalize, self-normalize, self-securitize, and indeed self-terrorize. They must mobilize themselves against the possibility that their own conducts, movements, gestures, reactions, or even instinctive responses will be what will lead to contemporary society's insecurity and terror. They must immunize themselves (by putting their flesh in danger as much as protecting it, as Esposito suggested[69]) against the ever present (virtual) possibility that they will be the responsible "agents" for the "ill health" of humanity itself, or what is left of it.

Ultimately, then, it is against their own self-induced terror that human bodies must guard. To guard against such a terror that one's being (mind, body, and soul) promotes, radical measures need to be in place, irrespective of whether the body remains alive or not. Human bodies thus end up conspiring in many contemporary situations of terror, whether they are conscious of it or not. In a way, without humanity itself being directly implicated in its own operations of horrific disfiguration, contemporary systems of terror and indeed horror would probably not be so effective. This is once again the crucial but terrifying insight one is led to draw from the kind of analytical perspective (introduced by Cavarero, as we saw in the "Introduction" to this book[70]) that recognizes the passage from biopolitical regimes of terror and terror reproduction to agonal conditions of horror.

A telling example of this self-mobilization and self-immunization against one's body and flesh can be found in the way Western states (or, rather, some of their governmental agents and agencies) along with some transnational organizations (the World Health Organization, the United Nations) asked populations the world over to preemptively take charge of their health, hygiene, and everyday routines in the context of the recent A/H1N1 (or "swine flu") pandemic. In this popular health scare, as with many other instances of spreading epidemics over the past decade (SARS, the H5N1 "bird flu," but also AIDS before), individual beings were asked to be the primary layer of securitization and defense against terror by turning their bodies (and those of family members, neighbors, co-workers, and so on) into primordial sites of scrutiny, intervention, and indeed governance. With the "swine flu" fear, a constant questioning of one's body movements and symptomatic features, along with one's daily habits, became an automatic (and auto-immune) and mostly unquestioned safeguard against the epidemic. Humanity's bodies, the physiology of the human species, but also the meaning of what it means to be human (and whether humanity should be reduced to a management of the flesh) became the main targets of the pandemic terror but, just as crucially, of the containment terror that followed the revelation of the pandemic. This governance of the "swine flu" and its scare (the disease and its terror were indeed inseparable from the moment a pandemic discourse was launched) was then based on a series of menial operations at the level of the human body and flesh (do I have a fever? Is my cough a sign that I have been infected? Did I remember to wash my hands after riding the bus or the subway?). Of course, it also relied on several self-cancellation measures (I must stay home for days if I feel sick; I must wear a protective mask if I venture outside and have a runny nose; we must close entire schools for as long as necessary if we suspect that children in the community have the flu).

In the end, the swine flu terror also raised vital questions about what humanity today is by proliferating a succession of moral quandaries that actually sought to reconfigure humanity as that which is always already primed for sacrifice. Among such so-called fundamental social dilemmas were questions like: Must I sacrifice my body and my self if my children are not yet sick but I am (and thus, I could infect them)? Will I have to allow myself to go without treatment (and possibly not make it) if it turns out that, as was reported, there are not enough vaccines for everybody? Must I segregate myself, and remove my body from the community of the living, so as not to let the remaining healthy bodies be even more at risk? The sacrificial logic involved in the swine flu scare may evoke some of René Girard's theorizations, particularly to the extent that this form of sacrifice appears to be aimed at the safeguarding of the community. But today's self-sacrificial or self-effacing logic goes far beyond the claim advanced by Girard that "the sacrifice serves to protect the entire community from its own violence."[71] Or, for that matter, that the ritual of the sacrifice "prompts the entire community to choose victims outside itself."[72] To the contrary, the encouraged sacrificial ritual of the swine flu pandemic terror is, to repeat, an auto-immune response or attitude that, far from casting violence away, actually seeks to keep it within and strives to spread it, sometimes in a horrific fashion, to as many living members of the community as possible. In fact, by placing itself beyond the threshold of both life and death, and by questioning the meaning of the human itself (and suggesting that the human should be redefined as that which is always ready for sacrifice), the swine flu sacrificial terror enters the realm of agonal politics and brings in the specter of a horror whereby humans must allow themselves to be quarantined, potentially culled, possibly manipulated and tested upon beyond death, and eventually left to decay in a condition of neither life nor death (not *bios*, not *zoe*, and not even *thanatos*) where the human is rendered unrecognizable as human.

From this perspective on how human bodies in societies of unease enable and reproduce conditions of terror that can open onto sites of horror, no centralized model of power can make sense of the fear that is necessary to these regimes of agonal or agonizing governance. At the same time, what these regimes of perpetuation of self-governmentalized terror tactics and bio-political insecurity call for is the beginning of a different understanding of what human life under agonal conditions means (as we already suggested in Chapter 1). It is not enough anymore to think of life as docile or regulated. It may also not be sufficient to think of today's living bodies as abandoned beings (as Agamben affirms[73]) who are caught in a state of endless sovereign exception (we will return to the issues of states and spaces of exception in Chapter 3). Rather, the self-rationalizing, self-securitizing, self-terrorizing, and self-horrorizing bodies and lives that act, react, and interact in societies of fear production today are closer to another analytical paradigm, one that was recently introduced by Mick Dillon and can be called "emergent life."[74]

For Dillon, emergent life is found in societies "governed by terror [and] in the process of trying to bring terror within the orbit of their political rationalities and governmental technologies."[75] Emergent life can be understood as a constant potential for adaptation or mutation inside terror systems. As Dillon clarifies: "[e]mergent means that they [living things] are capable of moving out of phase with themselves and becoming other than what they were."[76] The "living things" (as Dillon puts it) that constitute emergent life have no choice but to rely on pure contingency. This means that they can and, in fact, have to redefine their humanity incessantly inside the machineries of agonal governance that, ironically and cruelly, they also constantly reproduce. Because terror *dispositifs* are the environments that allow them to thrive, emergent human lives move, live, and exist. But they always remain on the *qui vive* or on the lookout for new ways of effacing themselves or denying their living/being as human. This emergent or auto-immunized human life is a life that continuously must watch itself and monitor its own movements, gestures, thoughts, beliefs, and feelings (as we suggested above) so that it can protect the system (not humanity) against its own self-induced dangers.

Dillon maintains that it is through the production, care, and destruction of emergent living/being that terror, unease, or insecurity is being governed, by which he means that it is placed within the domain of a certain promotion of life. As Dillon puts it, "the more effort that is put into governing terror, the more terror comes to govern the governors."[77] However, this promotion of life is, once again, beyond biopolitics. It is about inventing a way of being human (or inhuman since the distinction has become meaningless) that is always already located beyond life and death and, as such, is amenable to a violence and a horror that use and abuse the flesh of emergent life forms.

Among other things, what Dillon's thought on emergent living/being indicates is that it is time to push Foucault's thought on the biopolitical production of fear much further, perhaps beyond its biopolitical confines. For when we (and others) intimate the presence of a biopolitical productivity of fear or terror today, what we are pointing to is the existence of *a fear of fear itself*, or of *a fear of being fearful*. Docile and normalized bodies of biopolitical and governmentality regimes are not just afraid of not being able to live their normal life, as we hinted at above. They are also to be seen as emergent living forms that are designed to *fear being afraid* of living a life that has fear/terror as its vital impulse but also that are incapable of escaping such a terror. This is yet another dimension of the horror that awaits emergent "living things" as they are fixed or frozen by a fear of being afraid that, once again, allows them to be anticipatory and on the *qui vive*, but also prevents them from moving away from such a condition (here, we can recall Cavarero's useful distinction between terror and its capacity to put bodies in motion and horror and its paralyzing effects, as we mentioned in the "Introduction"). As we saw with the "swine flu" case, emergent humans fear being afraid not so much of the spreading disease and its social and physiological effects. Rather, they fear the terror that the disease (or any other danger) comes to represent. But this fear of the terror itself is unavoidable and, in a way, desirable or required for emergent life. By treating the pandemic (or the weather catastrophe, or the terrorist attack, or the nuclear scare, and so on) no longer as a possible natural or man-made disaster but as terror itself, a terror that, as Cavarero has argued, envelops one in fear but also opens up the door for horrific violence,[78] emergent living things deprive themselves of any possible solution or any resistant technology of living or being human that, perhaps, could tackle the problem that is said to be at the source of the terror (the so-called danger, although one should wonder whether such a danger matters at all as any encounter or circumstance in the life of emergent beings appears to be amenable to being the next terror).[79] Instead, the only way for emergent living things to deal with the impending doom is to fear more and more, that is to say, to produce more and more terror situations that will end up proliferating even more self-monitoring, self-carceralizing, and self-effacing techniques and *dispositifs* that, in turn, will confirm that they indeed had good reasons to be fearful in the first place. For today's emergent humanity, there is indeed nothing to fear but fear itself.

3 The *nomos* of exception and the virtuality of geopolitical space

> The solid ground of the earth is delineated by fences, enclosures, boundaries, walls, houses, and *other constructs*. Then, the orders and orientations of human social life become apparent.
>
> Carl Schmitt[1]

> The state of exception itself is thus essentially unlocalizable.
>
> Giorgio Agamben[2]

Introduction

How does a state of exception become normalized? How is space a prerequisite for the condition of abandoned being or bare life? How is the banishment of everyday life realized? What configurations of territoriality and power are at work in operations of abandonment? As we started to see in Chapter 1, these questions are among some of the queries posed by Italian philosopher Giorgio Agamben in his reflections on sovereign exceptionalism and its effects on human bodies/lives. Although asked by Agamben, these questions also assume an understanding of politics, of relations of power and force in particular, derived from Carl Schmitt's political thought (something we showed in Chapter 2). As several contemporary scholars have noted, these interrogations suggest that Schmitt's thought is far from being settled. In fact, Schmitt's work appears to be an endless source of creative investigation, a source that, of late, has enabled a series of critical explorations in geopolitics and international relations, particularly in the context of the Global War on Terror.

Although the general framework of this chapter is interested in Schmitt's writings on sovereign exceptionalism or, more precisely, on the biopolitical limitations of Schmitt's sovereign exception, the main objective here is to push another (perhaps less studied) aspect of Schmitt's work to the forefront of contemporary critical analyses (and to assess some analytical insights from Agamben's use of Schmitt along the way). Indeed, it is not just the frequently commented upon Schmitt of *Political Theology* or of *The Concept of the*

Political that concerns us in this chapter (unlike in Chapter 1), but rather some of the later works by Schmitt, particularly *The Nomos of the Earth*, recently translated into English (2003). There has been a beginning of a re-appreciation for Schmitt's *Nomos* of late, in particular in the wake of Schmittian derived concerns for the notion of the enemy or for the possibility of political order when a new US-based global *nomos* of the earth offers itself as a new universalizing legal and political system. Interestingly, Agamben has not found much to say about Schmitt's *Nomos.* But several critical geopolitical scholars and international relations theorists have suggested that Schmitt's writings on the spatial appropriation, delineation, and ordering of "the solid ground of the earth" can serve as useful points of departure for a necessary challenge to contemporary configurations of geopolitical power, force, violence, and war.[3] In so doing, many of these scholars have found themselves developing critical analyses at the junction of biopolitics and geopolitics.[4]

The recent revisiting of Schmitt's *Nomos of the Earth* has left one crucial question unanswered. Often eager to confine themselves to Agamben's own readings of Schmitt and to their linkages to biopolitics, biopower, or the idea of biopolitical sovereignty, theorists who have rediscovered Schmitt's *Nomos* by and large have failed to ask if and how Agamben's texts on the state of exception, or what sometimes he refers to as the "space of the ban," could be placed in relation not so much to Schmitt's decisionistic political theory, but rather to his later explorations about the connection between law and spatial ordering. A central preoccupation of this chapter is the question of whether what can be called a *nomos* of exception is in the process of being established in the global geopolitical landscape, often through a succession of practices or measures of isolation, abandonment, or exclusion of human bodies (as proponents of a biopolitical approach have argued), but also through operations of terror that open onto conditions of horrific indifference whereby the distinction between life and death becomes irrelevant. To capture some of the specific traits of this contemporary *nomos* of exception, and to reflect on the possibility that banning spaces today may reside beyond biopolitical frames of representability, we suggest that another analytical dimension needs to be introduced into both Agamben's and Schmitt's analyses. By extension, this added dimension could benefit contemporary studies that rely on a Schmittian–Agambenian framework to explain exceptional practices. This other, perhaps supplementary, but possibly self-standing, notion or dimension is virtuality, or the manifestation of virtual space and virtual territorial ordering. We argue that virtuality is crucial to the way a *nomos* of exception functions today, seeks to occupy geopolitical space, and contributes to the deployment of instances of agonal sovereignty.

What we call the *nomos* of exception of the contemporary geopolitical condition takes place when a virtual indistinction between being (bodies, lives, populations, races, species) and space (of the state, of the social, of the global, maybe of the human) starts to be what is primarily at stake in daily operations of violence, war, and terror. Instead of worrying whether the political exception—that act or decision that suspends the law in order to preserve a political order—is a materializable, localizable, or physically identifiable geopolitical phenomenon, or instead of trying to figure out where and when the exception may or may not take place (something that too many analyses on the spatialities of exception influenced by biopolitical perspectives have turned their critical gazes to), another critical posture can be adopted. This posture consists in retaining the notions of the ban and the exception but also in pushing them to their conceptual and political limits so that their capacity to usher in a series of horrifying effects on humanity can be apprehended. Indeed, as Agamben rightly posits, the state of exception or the condition of the ban/abandonment remains "essentially unlocalizable."[5] Yet, what renders the space of the ban or the exception so "unlocalizable" remains untheorized. This chapter attempts to remedy this analytical lapse in contemporary critical geopolitical/biopolitical analyses by suggesting it is as virtuality that the *nomos* of exception makes itself available to us—to our apprehension and comprehension—as "essentially unlocalizable."

The exception and the virtual

To theorize this unlocalizable territory of the ban and its connection to contemporary scenes and spaces of horror and agony, we may want to start by taking to heart Schmitt's opening remarks in *The Nomos of the Earth* about an ordering of space that would enable a "bracketing of war" (at least, between major geopolitical units, i.e., sovereign states) because it fundamentally would rest on the "solid ground of the earth" and on its accompanying territorial delineations. Schmitt suggests that what inscribes the *nomos* (a term that, etymologically, refers to land-bound law and that Schmitt equates with the creation of a "concrete organization of a community"[6]) onto the earth is not just fences, borders, and other traditional physical techniques and instruments of appropriation and demarcation of land, but also what (rather strangely) he refers to as unspecified "other constructs." It is as if the lines of demarcation and appropriation that define the *nomos* of a particular epoch into the land (starting with Schmitt's preferred model of land-law organization, the famous *Jus Publicum Europaeum*) should somehow be open, from the very beginning of Schmitt's analysis, to the possibility of as yet undetermined, undefined, or presently undistinguishable but nonetheless real, "other" delineating "constructs." In an era when the geopolitical organization of the earth has made room for a series of exceptional and often unspecified spaces (the camp, the prison, the airport, the border, the police checkpoint, the war zone, and so on), spaces that are no longer either inside or outside and whose occupants are often left in what some have called "limbo" states,[7] it may be time to consider the *virtuality of space* as one of those durable, commonly experienced, and real, yet insufficiently noticed or conveniently ignored (perhaps because we also do not quite know what to make of them) "other constructs" that Schmitt was perhaps already pointing to as potential markers of a *nomos* to come. We will suggest below that several concepts in Schmitt's *Nomos of the Earth* leave open an understanding of space—of the physicality of the earth and its markings (in a word, of *geo-graphy*)—that constantly wavers between visible materiality and what may be called "the real potentiality" of land appropriation. This "real potentiality" of land capture by "other" non-immediately present "constructs" is better expressed, we believe, by referring to the notion of a virtual territoriality.

To repeat, the challenge here consists of suggesting that the *nomos* of exception, or the generalization of a (geo)political order defined around the idea and practice of the ban/abandonment, is a land-law organization that accommodates itself fairly well of a geopolitics of virtual (always potentially actual) territoriality. Put differently, to the extent that it appears to work through materially delineated spaces (again, the camp, the border line, a city under curfew, and so on) and through marked bodies (the immigrant, the terrorist, the absolute enemy, and so forth), the state of exception and the order it ushers in take place and produce political effects precisely because they operate as virtual events, as always already potentially active conditions of actualization for a political order or, more accurately, for a political terror and violence to come. This virtualization of space, being, and political becomings is the domain of experience, the level of experimentation, and perhaps the zone of a cognitive power where the exception acquires its *potential* as a meaningful geopolitical construct in a context of agonal sovereignty. The ban and the exception are virtualities of contemporary geopolitical terror and horror because they are at once, in the same decisionistic instant, more outside the law than the exteriority of the law itself, and yet they never present a threat to the law or the legal order, unlike the absolute enemy, for example (to the contrary, they reinforce the legal order by stepping outside of it and outside the absolute outside/enemy/other too; they give the law its perpetual force). Likewise, the ban and the exception are virtualities on the inside too because their practices and measures appear to navigate inside the so-called sovereign political and legal domain more freely than the law itself may ever allow.

Perhaps we could turn to a more concrete example to clarify what we have in mind by virtuality here. In the wake of Agamben's theoretical reflections, it has been said that the camp is the prototypical form of the space of the ban or of exceptional territoriality.[8] We would argue that this may be the case, but only because the camp has become a virtual space or virtual condition. It is as a territorial virtuality that the camp becomes a paradigm of exceptionality and abandonment. The camp matters in contemporary geopolitics of exception or *as* the exception because it can, potentially, be actualized anywhere and at any moment. The camp is indeed exceptional because its geography is not fixed. Its geography does not and cannot limit itself to bodies or lives in space. For example, a prisoners' camp can be created on the edge (or in the center) of a metropolis overnight if a state of emergency requires it. A camp of quarantined flu victims or of wandering hurricane survivors can take hold of any public or private space at any time if a pandemic or a curfew has been declared (as we saw in New Orleans after hurricane Katrina where the Superdome was made into an instantaneous zone of triage for many of that city's flood victims). Likewise, the occupants of the contemporary camp are never fully determined in advance. Anybody under a condition of exception can find himself or herself on the "wrong side" of the law and interned in some prison, barrack, hotel, stadium, open field, and so on, irrespective of any nationality (as we discovered once it was made public that the occupants of Camp Delta in Guantanamo came from all over the world). Thus, as Cavarero has already intimated, it is perhaps humanity itself, the idea of the human, that is the general object of the virtual camp. In fact, it is not so much that the camp can be virtually anywhere. It is that it is always already everywhere.

It is important to understand what we mean by virtual territoriality in this context. By insisting on the virtuality of the *nomos* of exception, we are not claiming that we want to reject material geographies that point to the existence (in real time, with visible evidence, such as maps or photos perhaps) of "actual" spaces of exception (like Guantanamo or the Superdome in New Orleans). Our claim about the virtuality of the space of exception/ban is also not a defense of the kind of superficial conclusions that can be found in analyses for which the virtual today is merely synonymous with the false, the fake, the non-truthful, or the adulterated (as if there were a "good" order of representation as opposed to a "bad" or falsified one).[9] Moreover, virtuality in this analytical context also does not signify that the reality that is potentially present is somehow fictional, imaginary, or utopian. The possibility of virtual spaces (like the camp, as we saw above) is real, always in a process of sustained realization. As Philippe Quéau has noted, virtual representations or realizations have the capacity to "erase the boundaries between what we call the 'real' and what we thought was never a part of it."[10] Consequently, virtuality becomes a foundation for reality or, rather, as Quéau further clarifies, for what "we feel reality is."[11]

Just as importantly, the idea that virtual territoriality is always in a state of realization reveals a temporal dimension to what is often taken to be a strictly spatial problem of territorial ordering (i.e., the where as opposed to the when). Here, Gilles Deleuze's conception of the virtual offers us an important analytical supplement. Following from Henri Bergson's discussion of memory,[12] Deleuze's notion of the virtual is predicated on a conception of temporality that understands time as conditioned by what Deleuze calls an "immanent pure past," that is to say, a past that presents innumerable potentialities for future actualizations. For Deleuze, then, what is virtual is different from what is possible, or at least from what the possible often ends up signifying to most people. Whereas the possible is typically attached to the form or image of a "not yet" realized but still realizable representation, the virtual for Deleuze is inherently "real."[13] As Deleuze puts it, "*the virtual is real in so far as it is virtual.*"[14] To clarify further, the virtual is "real" because for Deleuze it is a largely non-corporeal form of presence, representation, or being that can only come to be as an event that, in its emergence, creates a present or marks a moment of actualization. This temporal relation between a pure past as virtual and the actualization of the virtual in the present is also what gives time (the time of the virtual) its space, even though the space of the virtual for Deleuze is not simply the product of a succession of momentary "nows" (or eventual actualizations). The space of the virtual, or virtual territoriality, is thus the entire temporality of the duration from pure past to present. The duration or, better yet, the spacing between the actual and the virtual past is what gives virtual territoriality a time and a space. More than the virtual or the spacing/timing from virtualization to actualization, Deleuze actually speaks in terms of an event. As Deleuze writes:

> The event is not the state of affairs. It is actualized in a state of affairs, in a body, in a lived, but it has a shadowy and secret part that is continually subtracted from or added to its actualization: in contrast with the state of affairs, it neither begins nor ends but has gained or kept the infinite movement to which it gives consistency. It is the virtual that is distinct from the actual … it is virtual that is real without being actual, ideal without being abstract.[15]

The temporal character of the Deleuzian virtual, and in particular its distinction from what is possible, helps us to comprehend how any *nomos* can be predicated on innumerable "virtual" spatialities that can be (maybe are waiting to be) actualized in specific (geo)political configurations, as events. Supplementing Quéau's more spatial understanding of the virtual with Deleuze's "time of the virtual" allows us to state that the virtuality of exceptional geographies is endlessly and boundlessly multiple, that "other territorial constructs" upon which spatialities of exception rely are mobile and malleable, and that they reflect emergent political temporalities that do not need referents or anchoring points in so-called material reality (or what Deleuze would call a "state of affairs") or in visibly bounded space (what one may want to see, locate, apprehend, or fix somewhere) in order to be forcefully deployed or meaningfully expressed and actualized. As suggested above, we encounter the virtuality of banning spaces as neither completely here with us nor totally removed from us, as neither relevant to us nor absent from our lives. In other words, banning spaces are always being actualized. As critical international relations theorist Nick Vaughan-Williams has astutely noted, this is perhaps why one of Agamben's most crucial reflections was that "we are all *virtually* homines sacri."[16]

Thus, the space of the ban encapsulates all of us, potentially, "really possibly," virtually, into a limitless domain of exception where life (the life of the human) can at any moment be relegated to abandon or indifference (and not even subjected to the force of the law or to the wrath of the sovereign, as biopolitical analyses continue to mention). This is one of the possible meanings of Agamben's claim that "we are all *virtually* homines sacri." Virtually, we are already in the space of the ban, pre-selected targets of a terror or horror that does not care if we live or die (such is life/humanity in the camp).

Following in the footsteps of both Agamben and Walter Benjamin,[17] Hardt and Negri have remarked that, in contemporary systems of management of terror, the sovereign decision on the exception has become a permanent condition.[18] The exception has become the rule. Once again, this insight has led other contemporary chroniclers of the condition of exception to be on the lookout for exceptional or banning "states of affairs," for zones of indistinction between the law and its outside that must be visible "now" and identified (or for various instantiations of the camp to be made visible, as if visible fixity was the irrefutable proof of the force of the contemporary virtual *nomos*).[19] While those studies have found it crucial to reveal areas where exceptional force primarily is exerted, what has been lost in these analyses is the recognition that the passage from exceptionality as a temporary event to the exceptional condition as a rule of permanence (or as an always potential present) is possible precisely because virtuality is the operative temporal and spatial mode of the ban/exception. To be sure, the prototypical exceptional event and its territorial dimensions may be more difficult to pinpoint when virtuality is understood as the condition of potentiality for the exception (or, as we suggested above, the camp must be revisited as being always already everywhere, always already built and ready to accept its human occupants). Thus, in addition to Guantanamo or Abu Ghraib (common geopolitical expressions and locations of the *nomos* of exception in the Global War on Terror), we may need to think about the London Tube,[20] or about most British cities and towns where CCTV cameras can be found at almost every street corner. In addition to the hapless victims of the Nazi concentration camps, we may also think about the Kurds in Turkish society who commonly confront a law that appears to exist only to them and for them,[21] or about Mexican migrant or undocumented laborers inside the United States who experience the sovereign exception not only by crossing the border but by having the border follow them through the joining of immigration policies, local police enforcement techniques, and accountability, safety, and productivity standards in the workplace.[22] In addition to the profiled individuals or the targeted ethnic groups after 9/11 in the United States, or after the March 11 attacks in Madrid, or again after the terrorist attacks of July 7 in London, we may also think about a college student in America or Europe (or anywhere else for that matter) who is doing research online for a term paper on the War on Terror, Al Qaeda, the concept of enemy combatant, or US and British militarism in Iraq after 2003, and who later finds out that her library holdings have been examined by some unspecified non-library entity.

The everydayness, the banal eventness, the mundane character of the *nomos* of the ban, of perhaps what Didier Bigo gestured towards when he coined the term "banopticon" (as we discussed in Chapter 2),[23] suggests that the so-called materiality of geographies of exception needs to be open to another domain of realization of time and space, to another experience of and with the real, that of the virtual. Critical analyses of geopolitics as biopolitics of exception thus need to account for and provide accounts of the ways virtuality—neither here, nor there, yet both at the same time; neither you, nor me, yet potentially all of us already—constantly makes, unmakes, and remakes real bodies and actual territories. Such a posture, as we suggest in the next section, is not antithetical to Schmitt. Indeed, like Schmitt, it is possible for contemporary theorists of the virtuality of exceptional/banning spaces to chart the territories, borderlines, outer limits, and relations of "real possibility" (including relations of inclusion and exclusion, friendship and enmity) of the new *nomos* of exception.

Schmitt and virtual territoriality

Suggesting that Schmitt entertains the "real possibility" that "other constructs" could play a part in the making of a land-based legal order or *nomos* is probably not enough to assert that Schmitt was a political theorist of the virtual all along. In fact, the suggestion that Schmitt's conception of the *nomos* is amenable to virtual territoriality may encounter a common objection among Schmittian scholars. How can we argue that Schmitt opens the door for virtual territoriality when his thought about the *nomos* is supposed to be so earth-bound, so materially inscribed in the soil, so physically embedded in "mother Nature" or in what Mitchell Dean has called a "mythology of the earth"?[24] Does Schmitt's fundamental commitment to an engraved and occupied land as the ferment from which both *Ordnung* (order, law) and *Ortung* (location, place) emerge not preclude virtual spatiality and, by extension, the idea that a *nomos* of exception could be derived from his analyses?

After all, Schmitt's *Nomos of the Earth* is famous for its abundance of physical inscriptions, its earth/soil/land references, and its images and metaphors derived from the idea of a material rootedness of human existence. His thought about order and humanity has been described as "telluric."[25] Indeed, for Schmitt, the Greek word *nomos* refers first and foremost to a "concrete enclosed location (*Ortung*) on the surface of the earth."[26] As political theorist Mika Ojakangas suggests: "Schmitt time and again stresses that the true law has an intimate relationship with soil (*Boden*) and land (*Land*). It is always bound to the earth (*Erde*)."[27] For Schmitt, the inscription of a concrete order in the earth and the derivation of the law from the soil also explain why *delineation* (of terrains, territories, and fields, or what topography amounts to for Schmitt) and *appropriation* (of uncultivated or virgin lands, states, and resources, or what one might suggest political geography ought to be for Schmitt) are crucial concepts. As Schmitt affirms: "[s]oil that is cleared and worked by human hands manifests firm lines, whereby definite divisions become apparent."[28] Schmitt adds that "land-appropriation precedes the order that follows from it. It constitutes the original spatial order, the source of all further concrete order and all further law."[29] Thus, Schmitt appears to insist on a fundamentally geophysical character of legal order and political organization. It is in fact the connection to the earth that makes any order and organization concrete. Ojakangas intimates that land, soil, and earth for Schmitt are expressions of a form of "primevalism,"[30] one that consists of wanting to ground all human (political, social, legal) constructions into *Ur* (or primal, primeval) concepts. Schmitt himself writes that "the great primeval acts of law [are] terrestrial orientations."[31] Such a geopolitical "primevalism" necessarily determines (grounds and bounds) anything, any object, any condition of human possibility, and any life and meaning. From this perspective, it may indeed be hard to fathom that virtual spaces (such as the camp, as we saw above) could be derived from such a foundational attachment of all things human to the earth.

Still, upon closer examination, Schmitt's text does offer some openings that take him away from such a foreclosed physical "primevalism." In fact, we want to suggest that the condition of possibility for the virtual (and virtual territoriality) is present in Schmitt's *Nomos of the Earth* in some passages that, at times, appear to be mere additions, supplements, or clarifications of some of Schmitt's other concepts about the earth-bound encasement of order and space. One such key passage deals with the issue of the nomad and its relationship to land. Interestingly, this passage can be found in an appendix (itself, a mode of supplementation) to the *Nomos of the Earth*.[32] Schmitt insists that the nomad too has his own *nomos*, his own way of delineating and appropriating open land and space, even if it is a rather primitive and incomplete *nomos* when compared with the modern inter-state legal order of the *Jus Publicum Europaeum*.

The nomad (and the etymology of the term seems to confirm it) is not free to roam the earth boundlessly and cannot escape the material conditions the soil/land offers. The nomad occupies a space too, appropriates lands, and inscribes his/her order/law onto the territory where it travels. At the same time, Schmitt is forced to recognize that the nomad does not cultivate the land. The nomad is more like a shepherd (*nomeus* in Latin) who "tends to" or "looks after."[33] According to Schmitt, (s)he does not cultivate or build like a patriarch or a statesman would. Moreover, the space the nomad occupies constantly shifts. As Dean suggests, the nomad for Schmitt "wanders in search of pasture."[34] This leads Schmitt to state that the land-appropriation of the nomad is based on "perennially provisional appropriations and divisions."[35] While the nomad moves on and throughout the earth (thus, walks onto the soil and leaves his/her marks into it), the territory of the nomad has expandable boundaries. This is perhaps why the nomadic *nomos* is a transitional one for Schmitt, or a "prelude" to modern territoriality.[36] Containment and location, grounding and fixing are anathema to nomadic life. Travel, movement, and displacement are instead conditions of possibility for the nomad not only to live, but also to have a space, itinerant and shifting as that space may be. One could almost say that travel, movement, and displacement are pre-conditions for the space/territory of the nomad. They pre-exist and pre-determine what the nomad will make of his/her space, how (s)he will organize it, where and how far (s)he will go, and what forms of political, social, and economic life (s)he will be able to establish. As Gilles Deleuze and Felix Guattari have claimed, nomadism is about making use of and proliferating (through one's actions and motions) "smooth space."[37] It is not about fixing oneself into and recognizing as sacred and fixed the "striated space" of the law and state power.

Clearly, as Schmitt recognized, the nomad has a space and a *nomos*, perhaps an infinite expanse of the earth's territory to occupy. But it is not clear how this mode of occupation of land and space is equivalent to that type of sedentary occupation/appropriation of soil-based places and activities Schmitt takes to be the foundation of the mostly European geopolitical *nomos*. While Schmitt admits that the *nomos* of the nomad is transitional or provisional, Schmitt still finds in it a basic modality of delineation and appropriation that qualifies it as earth-bound. But, as Schmitt is forced to recognize, the nomad's own delineation–appropriation–occupation sequence is not based on cultivation since cultivation requires fixity. Moreover, the dimension of delineation of the nomad's *nomos* is particularly confusing. What lines does the nomad's wanderings trace? Can they be clearly established, instituted, or defended? Perhaps, as Deleuze and Guattari have intimated, we may be better off thinking of the nomad's tracings onto the earth as "rhizomatic" (as opposed to arborescent, or tree-like).[38] The lines of the nomad are not like the early modern *rayas* or the national sovereign states' borderlines that Schmitt is more likely to associate with the *nomos* of the earth. It is not even obvious that they could count as precursors to such lines. Rather, they are markings or footsteps that both follow and precede the nomad's constant movements and displacements. They have no apparent beginning or end point. They do not easily delineate a here and a there, an inside and an outside, or an us and a them.

We want to suggest that, to the extent that Schmitt's nomad does have a *nomos*, it is a virtual one, a *nomos* premised upon the "real possibility" of virtual territoriality. The lines in the ground that Schmitt is looking for and claims are the necessary foundations for any proper land-law order and any organization of a human community are not imaginary, however, in the case of the nomad. They are there; they exist; they have an actuality. But they are also different. And it may not be enough to refer to them as "provisional," as Schmitt does (as if those nomadic markings were dotted-lines whose blank spaces early modern *rayas* and sovereign boundary-lines would later fill). Provisionality would suggest that nomadic lines are simply "not yet" sovereign border lines (as Deleuze's theorization of the virtual would intimate). As suggested above, these nomadic lines are trajectories that may or may not successfully connect the daily meanderings of nomadic existence. More than lines in the soil, markings onto the surface of the earth, they are memory traces, souvenirs/memorizations (and their narrations) of past travels as well as anticipations in both time and space about journeys that may or may not chart a territory ahead. These tracings or footsteps that, once again, follow and precede the nomad and his/her ceaseless movements are lines that do visualize and virtualize an actual space, the space of nomadic life. Still, they often do lack the traditionally earth-bound reference points or markers that normally demarcate modern geopolitical territories. Only as virtual (already passed through and always about to be actualized) is the territory of the nomad experienced. Only through the passing of time (a duration or spacing from pure past to actual present, Deleuze might say) as well as by way of travel and displacement can lines and appropriations of space be said to be meaningful to the so-called nomadic *nomos*. But those nomadic lines and appropriations are in a condition of virtuality, particularly vis-à-vis the earth-bound territory of Schmitt's preferred modern and European political *nomos* that enables the existence of sovereign nation-states (once again, Schmitt's preferred geopolitical "state of affairs").

Thus, Ojakangas is on target when he writes that, for Schmitt, "although every *nomos* implies an enclosure, every enclosure is not necessarily a *nomos. Something more is needed.*"[39] This "something more" that is "needed" is the understanding, perhaps already hinted at by Schmitt in his mention of "other constructs," that virtuality (of space, of order, of being, of time) may come to the rescue of the material *nomos* and facilitate its actualization when delineations and appropriations cannot be evidently shown to be grounded into the earth or the soil (as the example of the nomad above showed us). Thus, to Ojakangas' statement, one could add the following clauses: that every *nomos* is never an immediately recognizable and fixable enclosure; that it is an always already potential enclosure; and that it is to the virtual that, in some important measure, Schmitt's *nomos* owes its potentiality to operate as an enclosure for human life.

Another dimension of Schmitt's analysis of the *nomos* that seems to encourage the idea of virtual territoriality is also introduced by Ojakangas (although it must be said that Ojakangas is not interested in mobilizing the notion of the virtual in his studies). Ojakangas implies that Schmitt's notion of land appropriation and its connection to the creation of a concrete geopolitical order (in particular, the famous European legal order of territorially based and demarcated inter-state relations that prevailed until the end of the nineteenth century) is necessarily dependent upon the presence of what Schmitt calls a "space of the outside" or "exterior space" (*Ausland*).[40] This "space of the outside" is a space that must remain clearly separate from the inside (the state, the nation), but also, and perhaps more critically, must remain "open and fluid" or "fixed yet not ossified."[41] Thus, this *Ausland* is not just an "outside territory" (as opposed to an inside, or what must be contained within fixed boundaries). Rather, this exterior space is the condition of possibility for any orderly inside/outside division.

To better capture the peculiar meaning and place of this *Ausland* in Schmitt's designs about geopolitical space and order, we need to turn to Schmitt's well known fear about the dangers of a unification of the world into a single dominant religious, ideological, or political system.[42] The possibility that the pluralism crucial to the *nomos* of the earth (multiple sovereign states with their own lands safeguarded by borders) may give way—as in fact happened in the early twentieth century, according to Schmitt—to a new ungrounded and universalizing conceptualization of order is a terrifying prospect. As Schmitt already stated in *The Concept of the Political*: "the political world is a pluriverse, not a universe."[43] The transformation of a pluriverse of sovereign states into a universal order dominated by one vision of humanity (with democracy and human rights spreading throughout the world, for example) may well be the consecration of cosmopolitanism and a certain version of liberalism on a global scale. But, for Schmitt, it is also the terrifying certainty that wars without end will be endlessly waged on behalf of the defense of humanity. Ironically, only inhumanity can result from this desire to unify and defend humanity at all costs. Schmitt famously argued that two major consequences are to be expected from this imposition of universalism (or, at least, from this semblance of universality often masking the political, economic, and cultural imperialistic designs of a hegemonic power) and from the creation of a new de-territorializing *nomos*. First, the bracketing of war that was a characteristic of the *Jus Publicum Europaeum*—the possibility that wars may still take place between sovereign states, but in a limited and regulated fashion, or on the outside of the inter-state system (colonial wars, for example)—now gives way to a total prohibition of war.[44] Outlawing war, casting it out of the new *nomos*, actually ensures that multiple dehumanizing "wars to end all wars" will take place, thus rendering humanity ever more insecure and potentially bringing it closer to total destruction. Second, the concept of the enemy as a concrete public opponent, one with rights and guarantees and endowed with respect, disappears in favor of the notion of an absolute enemy, a totally terrorizing outlaw whose very presence becomes the mark of inhumanity itself.[45] Such an enemy cannot stand in the way of universal progress, or of global uniformity, and must be endlessly fought and destroyed. As Schmitt puts it, the collapse of the old European land-based *nomos* gives way to an ungrounded, unbound, and uniformizing system (from a pluriverse to a universe) according to which the principle of recognition of a "just enemy" is sacrificed to the requirement of the implementation of "just wars" fought on behalf of the defense and integrity of the liberal and democratic global order (from *justus hostis* to *bellum justum*).

The notion of land appropriation (*Landnahme*) is also totally excluded from the idea/reality of a new universal *nomos*. What is there to appropriate, cultivate, mark in the ground, or make one's own when the world is unified and a global law is invented and expanded to preserve a system of unification and uniformization (for example, by condemning the recourse to war by individual states to sort out their disputes and, instead, devising an international system of collective security—with global institutions and international courts—that will transcend such individual and selfish actions)?[46] The removal of land appropriation is, according to Schmitt, the prerequisite for the establishment of any universal order. Thus, as Ojakangas notes, if the *nomos* of the earth is to remain pluralistic, the "real possibility of appropriation" (Ojakangas' own phrasing) must be maintained.[47] Maintaining or, perhaps, reinventing a system of appropriation by multiple land-based political entities (although not necessarily nation-states; geopolitical blocs may have to do) is the only possible safeguard against the imposition of a universal order. To keep open the possibility of appropriation in the *nomos*, a "space of the outside" must thus be preserved. In other words, there always must be an uncharted zone that lies beyond, or remains in excess of, any delineations and possessions by political entities (even beyond the inside/outside divide). Otherwise, a system of total saturation of space, or a world unity, will be established.

The crucial dilemma encountered by Schmitt in the twentieth century is how to rediscover the possibility of the *Ausland*, of territories always open for capture, when all lands (and even the seas) have been colonized, are all occupied, but also now find themselves under the dominion of one global order. It is precisely at this point, we would argue, that virtual territoriality comes to Schmitt's rescue again. In another passage from the "Nomos—*Nahme*—Name" supplement/appendix to *The Nomos of the Earth*, Schmitt writes:

> Allegedly no longer is anything taken … An important representative of political science at a leading university in the United States recently wrote me: 'Land-appropriation is over and done with.' I replied it has become even more serious with the appropriation of space. We have no right to close our eyes on the problem of appropriation, and to refuse to think any more about it, because what one today calls world history in the West and the East is the history of the development in the objects, means, and forms of appropriation interpreted as progress. This development proceeds from the *land-appropriations* of nomadic and agrarian-feudal times to the *sea-appropriations* of the 16th to the 19th century, over the *industry-appropriations* of the industrial-technical age … , and, finally, to the *air-appropriations* and *space-appropriations* of the present.[48]

Here, and rather unexpectedly, the notion of land appropriation is expanded by Schmitt to other seemingly equivalent forms of appropriation (dictated by "progress," as he puts it). While they are supposed to be extensions of the land-based model of human possession of the soil/earth, these new appropriations are surprisingly ungrounded, or detached from "mother Earth." In fact, not only are sea and air appropriations accepted by Schmitt now, but even what he calls "industry-appropriations" and outer space-appropriations can count as territories in need of human capture, delineation, and productive exploitation. In an era when a universal order of the earth is about to prevail (if it does not already), these non-terrestrial spatialities must be urgently recruited. Otherwise, the "free space of action" that is guaranteed by the presence of the *Ausland* will forever disappear.[49] However, this *Ausland* made visible and possible through industrial or even celestial appropriations is only virtually akin to the *Ausland* of the old terrestrial *nomos*. Or, perhaps, it is the fact, now revealed by Schmitt, that the notion of a "space of the outside" was always intended to be in excess of land, physical territoriality, and the earth. This *Ausland* that guarantees the endless "real possibility" of territorial capture and human cultivation was perhaps always supposed to function as a virtual territorial construct in Schmitt's work, or as a concept that could virtually ensure the continuation of appropriations and delineations by humankind irrespective of whether there are any unoccupied, open, and free lands and soils (or even seas) to be found anymore. Put differently, the "really possible" extension of the *Ausland* through appropriations that are eminently de-territorial activities can give birth to a virtual *nomos*. This virtually concrete (always on its way to being actualized) ordering of space and people is nonetheless presented as the main challenge to an otherwise likely universalization and unification of the world. As Schmitt writes: "what would be terrifying is a world in which there no longer existed an exterior (*Ausland*), but only a homeland (*Inland*)."[50] The virtuality of space is thus for Schmitt an important way of fending off that terrifying outcome. Virtually appropriated territories make sure that the *nomos* of the earth (old or new) is actually and continuously re-territorialized.

In a recent but too brief account of how Agamben actually tries to borrow a few insights from Schmitt's *The Nomos of the Earth*, Dean argues that Agamben hopes to arrive at an appreciation for a new *nomos*, one that would be able to encompass our contemporary condition of terror or horror, abandonment of bodies, and endless war. The development of this new *nomos* suggested by Agamben is not just an extension of the universal liberal democratic system of law and order deployed in the name of humanity and policed mainly by the United States, which Schmitt uncovered as a terrifying possibility (and sought to react to). It is rather much closer to the notion of a *nomos* of exception or of a generalized space of the ban that we introduced earlier in this chapter. Dean suggests that, in wanting to connect Schmitt's notion of land-appropriation (*Landnahme*) with the idea of exception (*Ausnahme*), Agamben goes off track. Dean believes: "Agamben ignores the overall elemental framework in which the terrestrial character of the *nomos* is located."[51] Dean adds that the Agambenian linkage between *Landnahme* and *Ausnahme* ends up "enclosing the *nomos* in 'the camp' as to render the struggles over the rest of the planet … scarcely intelligible."[52]

The problem with Dean's reading of Agamben and, by extension, of the notion of a *nomos* of exception is that it ignores the role and place of virtual territoriality both in Schmitt's work and in contemporary practices of abandonment and exception. While it may be the case that Agamben retains a limited or fixed conception of the camp in his work, the notion of the camp—its physical contours and appearances, its occupants, its meanings under a politics of terror and horror—needs to be revisited when virtual territoriality is introduced as a condition of possibility for the contemporary *nomos*. The next section begins to undertake this investigation of the linkage between virtuality and the space of exception. But it does so *not* by claiming there is a seamless correspondence between the presence of virtual territoriality in Schmitt's work and Agamben's recognition that, under conditions of exception, "we are all *virtually* homines sacri." Indeed, it is not even clear that Agamben is aware of such a correspondence. Rather, we suggest that the condition of possibility for the exception and the ban today require a conception of space that can connect the construction of the *nomos*, potentially any *nomos* (and its appropriations, delineations, and productions), to the notion of an open, fluid, movable, changeable, transient, unfixed, and perhaps nomadic virtual territoriality. As was explained, virtual territoriality was already probably crucial to Schmitt's thought on the *nomos*. We believe that it is also crucial to Agamben's exposition of the state of exception and the space of the ban. We show in the next section that a few critical geopolitical theorists have started to reflect on the importance of virtual territoriality for today's *nomos* of exception, and we take advantage of their insights to connect the notion of the virtual to contemporary considerations of terror, horror, and agony.

Virtualizing terror and horror through the space of the exception

In a recent analysis of Guantanamo's Camp Delta as an allegedly prototypical contemporary site of exception, Derek Gregory sought to base his critical study of what, in this chapter, we have called the *nomos* of exception on both Schmittian and Agambenian theoretical foundations.[53] Drawing insights from the work of political theorist Mick Dillon,[54] Gregory posited that "conceptions of space need not be limited to the container model."[55] This suggestion not to limit oneself to the "container model" of territoriality is interesting because it echoes Schmitt's own invitation to consider the possibility that any solid and durable legal and political order may be reliant on "other constructs." Gregory's statement also parallels Agamben's claim that the state of exception is "essentially unlocalizable." As was suggested above, virtual spaces, expanses, or enclosures may be considered to be territorial constructs necessary to the execution of the exception. Places and conditions of life enacted through a virtuality of space may not always be immediately evident, visible, or even realized. But they are nonetheless concrete and actual geopolitical situations or events that often enable legal, political, and military institutions to establish an order of the earth, even if such an order is premised upon a banning of bodies and lives.

In addition to Quéau's and Deleuze's respective insights with regards to the virtual, Jean Baudrillard's theory of simulation has revealed that virtual territories matter because they often stand for the so-called real and can become more real and truthful than reality and truth themselves.[56] Importantly, one of Baudrillard's favorite images to theorize the hyperreality or virtuality of simulation is that of the relationship between the map and the territory. Baudrillard writes: "Simulation is no longer that of a territory, a referential being, or a substance. It is the generation by models of a real without origin or reality: a hyperreal. The territory no longer precedes the map, nor survives it. Henceforth, it is the map that precedes the territory—precession of simulacra—it is the map that engenders the territory."[57] While Deleuze may object to the claim that the "real" of virtuality (or its actualization) is "without origins" (as Baudrillard maintains), the key point for us is to recognize (with Baudrillard) that, as hyperreal constructs, virtual territorialities may indeed be our main guarantee that something which insists on being real in the political domain still matters. As we saw above, it appears that virtuality already served such a function for Schmitt.

Thus, when Gregory stated: "sites like Guantanamo Bay need to be seen not as paradigmatic spaces of political modernity … , but rather as *potential* spaces whose realisation is an occasion for political struggle,"[58] he may have been gesturing towards the idea that the contemporary camp, this "limbo zone" between life and death, can exist and produce meaningful political effects precisely because it operates as a virtual territorial construct. As we argued above, it is the always plausible actualization of camp-like conditions anywhere and anytime that gives the space of the ban its powerful political presence, its forceful and violent capacity to shape the contours of a generalized space/order of exception or *nomos*.[59] However, to now take this analytical logic one step further and to insert into it the perspective on the virtual offered by Baudrillard, what is also virtual about Guantanamo (or other camp-like sites) is the fact that the exceptional and banning potential of this space may have already been realized in many other places *not* officially recognized, described, or sanctioned as camps.[60] In other words, the virtual "real possibility" of the exception (Guantanamo, Abu Ghraib) may also seek to draw our attention towards the idea of a supposedly unique, unusual, extreme, or indeed "exceptional" zone *as if* such a site were *not* the norm, or *as if* it were *not* supposed to be real or actualized anywhere else. As with other simulated mechanisms or operations,[61] the virtual space of exception hopes to make it look like everywhere outside these supposedly punctual or fixed camps that are recognized to entrap and abandon bodies, the logic of the ban is actually not widespread or common whereas, once again, the exception may have already (virtually) become the rule.[62] There is thus a strategic (geo)political dimension to the deployment of virtual spaces that, unlike Deleuze's perspective, Baudrillard's conceptualization of the virtual makes us aware of. As we will argue towards the end of this chapter, such a strategic deployment of virtuality is important to contemporary instances of agonal terror and horror.

International relations theorist Nick Vaughan-Williams is one of the few critical scholars who has started to look into the relationship between strategies of virtuality and the contemporary politics of exception and abandonment. Among other things, Vaughan-Williams has shown how simulated exercises or "fake" gaming scenarios about border crossings into the United States and about controlling illegal immigration in Europe have been deployed to substantiate the meaning and reality (including the physical presence) of the border and of the territory (the inside) supposedly protected by it.[63] Through those simulated exercises, games, or testing programs, it is not just the point that the integrity of the territory to be secured is endlessly displaced or that geographies of border demarcation are increasingly "unlocalizable" (and thus can be actualized anywhere). Indeed, they are. But the point is also that the virtualization of space—the supplementary layers of a space to be secured that are intended to be more real, meaningful, and effective than the typically and traditionally understood geographical space of the state, the nation, or the sovereign order itself—inevitably proliferates the exception and its territoriality of abandonment everywhere.

Yet, though it spreads everywhere, the exception is not to be found in any particular place either, as we argued about the camp above. The space of the ban is thus virtually nowhere as well to the extent that bordering practices, even when simulated, endlessly recreate borders and border controls everywhere. But, by the same token, such a banning space virtually, potentially, and "really possibly" repositions bodies throughout the globally indeterminate order of the ban and, in so doing, ceaselessly regenerates possibly abandoned beings. Thus, unlike Dean who believes that Agamben remains tied to a fundamentally referential or representational geographical understanding of the *nomos* of exception (or of the camp as fixed onto the earth), Vaughan-Williams intimates that a virtual logic of spatiality is present in Agamben's analyses. As Vaughan-Williams puts it:

> Agamben's diagnosis is highly suggestive of a new kind of *nomos* of the earth … , one that reveals a far more complex, and less spatially and temporally fixed, relation between juridical-political order and spatial orientation. As hinted at in Agamben's italicized use of the word "virtual" in his prognosis "we are all *virtually* homines sacri," this re-worked *nomos* is simulated throughout everyday life in such a way that potentially emplaces us all under conditions of considerable uncertainty. On this reading, the sovereign decision to produce some life as bare life is not one that necessarily happens literally at specific points in space and time … Rather, in the order of the virtual as described by Baudrillard, this decision can be re-read as a far more generalized and reiterative process in the attempt to simulate total security.[64]

One conclusion that can be drawn from Vaughan-Williams' rereading of Agamben's space of the ban (with the help of Baudrillard, in particular) is that the distinction between real spaces and virtual spaces of abandonment is no longer relevant, perhaps just like the distinction between real and virtual land appropriations eventually became meaningless for Schmitt. This distinction no longer matters because real spaces of the ban owe their geopolitical relevance to a virtual condition opened up by the necessarily extra-legal potentiality of the juridico-political order (the law must—potentially, virtually—always be in excess of itself to maintain and reproduce the legal order). Virtual spatialities contain, capture, include and exclude, and banish as much as so-called real boundary lines or real territorial appropriations. Perhaps political philosopher Steven DeCaroli says it best when he announces that "a necessary condition for the possibility of banishment is a boundary—*real or virtual*, terrestrial or divine."[65] For DeCaroli, the distinction between the reality and the virtuality of the *nomos* of exception is fundamentally inconsequential, indeterminate, and probably undecidable. Or, to return to Deleuze, it is always the case that a real phenomenon owes its actuality to its virtual possibility (the plural possibility, or multiplicity, of its emergence as an event).

But if it is correct to argue (as many critical geopolitical scholars have done lately) that the condition of exception is fundamentally geographical,[66] what does it mean for our understanding of the space of the ban in an age of horror to state that the distinction between real and virtual spaces is always already erased and meaningless? What new critical analytical horizons does this connection between virtuality and the exception allow us to perceive? As a point of departure to answer these questions, the recent work of one critical geopolitical scholar (also an important reader of Agamben), Claudio Minca, needs to be examined closely. Indeed, Minca offers us some important indications as to what sort of geopolitical reflections are likely to emerge as a result of the recognition of the connection between the virtual and the exception. Thus, Minca's texts provide us with an intriguing conceptual platform from which our own analysis of the relationship between virtuality and agony can spring.

At first glance, Minca appears to uphold the belief that the *nomos* of exception is eminently terrestrial, geophysical, and perhaps "primevally" material. Similar to previous instantiations of the *nomos*, the order of a generalized exception/ban seems to be derived from an inscription into "Mother Earth." In a typically Schmittian fashion, it is the manifestation of an ordered essence that one can see, locate, apprehend, or even use for the purpose of political divisions. Or, as Minca puts it (in a language that cannot fail to recall Schmitt's "solid ground of the earth" imagery), the space of exception is a zone of indistinction between life and death that "must be somehow 'traced on the ground' to materialise, to allow violence to be effective; to allow for the exercise of a power with no references or persuasion."[67] The decisionistic power that marks the exception (in both Schmittian and Agambenian theories) is a power that, as Minca further notes, must find a "material and mappable space" in order to become effective.[68] Still, it is a strangely localizable or fixable space (notwithstanding the fact that Agamben already referred to this space as unlocalizable), a rather odd and shaky ground, and certainly a blurry and confusing map that are all required for this endless going back and forth between life and death, between the law and its outside, to be realized. In fact, Minca readily recognizes the unusual amorphous and disturbing nature of this "ground" that, he believes, is a precondition for the condition of exception and abandonment. He refers to it as "an unsettling point of contact."[69] In another essay, he calls it the "arcanum" or the "secret coordinates" of a geographical non-space.[70] In doing so, Minca may not go much further than Agamben who referred to the space of the ban as a "dislocating localization" but never theorized its geopolitical structure beyond this statement.

At the same time, there is quite a bit of an extension or improvement on Agamben's analyses to be found in Minca's work. Minca provides us with conceptual tools that can help us to decipher the alleged secret enigma (or "arcanum") of the zone of indistinction that renders the exception so common, normalized, and readily actualized. To be more precise, it can be argued that Minca offers an analysis of banning spaces *as* virtual spaces. In fact, what Minca takes to be the "arcanum" of the condition of exception is virtual territoriality. Understood as the virtual, the secret of such a space becomes far less mysterious. Indeed, when Minca declares that "the biopolitics of exception … insinuates itself insidiously into our everyday lives, invisible, the product of a space of indistinction, a space that is neither inside nor outside,"[71] he is already talking about virtuality. At least, Minca partially mobilizes the language of the virtual in order to make his point about the state/space of exception. Perhaps he only refrains from openly talking about virtual territoriality in this context because of what may be some misconceptions (not uncommon among critical international relations and geopolitical theorists, as we stated above) about what the virtual is and does. For if it may be understood that virtual spaces (virtual enclosures, virtual borders, virtual lines, and so on) have a somewhat "invisible" (as Minca puts it) or perhaps non-immediately materializable but nevertheless very concrete and indeed mundane quality about them (since this spatiality "insinuates itself in our everyday lives," says Minca), to then go on to affirm that the space of the exception is "insidious" is something that is not logically derived from the claim (that Minca also seeks to uphold) that "the exception everywhere becomes the rule."[72] A space and a condition of abandonment that become so "exceptionally" common and readily propagated only remain a mystery, an enigma, a secret machination, or an insidious undertaking (by some orderly biopower or biopolitical force, perhaps) *if and only if* one wishes for them to remain enigmatic, mysterious, or secret. This veil of secrecy, this "arcanum," like a Baudrillardian simulation in fact, is what allows one not to face the "reality" of virtual territorial constructs, what permits one to continue to declare that the *nomos* of exception must be "traced on the ground." The preservation of this "arcanum" may thus reflect a nostalgia for the logic of the geopolitically or biopolitically real, material, and representational (as if there had to be a one-to-one relation of equivalence between an object and a subject, or a map and a territory). But we would warn here that one maintains the space of exception as a secret enigma at the cost of apprehending the condition of "real possibility" of a political order that potentially, always already, has turned us all into *homines sacri*.

Revealingly, Minca's analyses actually go back and forth between a reading of power (or *potenza*, as he calls it) as geophysical and biopolitical materiality, with "real" effects on bodies and lives in "mappable space," and a theorization of power/potentiality/*potenza* as virtuality. When Minca writes that the power of the exception "finds its most potent expression *when it becomes spatialized*; that is when it materializes the condition of the suspension of 'normal' order within determinate spaces, spaces where, literally, *everything is possible*, and everything can and does happen,"[73] the spatial *potenza* he invokes here is virtual territoriality, virtual bordering, virtual securitization, or virtual dehumanizing violence. It is, as Minca puts it so lucidly, the force, the power, the potentiality, the virtuality (vir-tuality comes from *virtus*, or force/energy in Latin[74]) of a "true *potenza* … that needs neither a before nor an after to justify itself, but that manifests itself as such."[75]

This *potenza*/power does not require a logical temporal progression from a nothing to a not yet, and all the way to a now and a here that would mark a series of mappable spaces and places of exception. What it requires, rather, is a gap or a spacing between what Deleuze once again referred to as a pure past and an actualization as event. Thus, the *potenza*/potential for an all-encompassing or all-targeting violent dehumanization across virtual camps is always in existence or in motion (or we are always in its presence), but it only acquires the status of event and gets recognized as such when it manifests itself. Put differently, the virtual force/power of the exception is that it actualizes itself in the very instant/event that nonetheless does not resolve or saturate its potentiality or "real possibility" (in this fashion, it recalls the "possible eventuality" of war that, according to Derrida, could be found in Schmitt's work, as we saw at the end of Chapter 1). In the exception and as the exception, the real then is of the order of the virtual, and the virtual is always already real. And, as Minca, Gregory, and others have concluded, in this sort of geopolitical *nomos* where real spaces are "really potentially" shaped and structured by virtual emplacements, everything everywhere and every time "is truly possible" for everybody.[76] Thus, to repeat what we claimed at the onset of this chapter, identifying sites of exception or singling out bodies to be placed in camps while others, allegedly situated outside the exceptional space, will remain safe and protected is a futile political and analytical exercise. As we stated in the "Introduction" to this volume, it is an exercise that makes sense only if one takes for granted the mode of representability and intelligibility advanced and privileged by proponents of a biopolitical approach.

Conclusion

The possibility of a "true *potenza*" mentioned by Minca above recalls the "real possibility" of the preservation of a "space of the outside" or *Ausland* that was discussed by Ojakangas in relation to Schmitt's concern with land-appropriation. It is this "true" or "real" possibility for humanity to experience the space of the ban, the violence of the exception, the terror of a law that is neither here nor there, and the horror of an order that (as Agamben still persuasively argues) does not punish but rather banishes, that normalizes the *nomos* of exception. It is the virtual force/power, the *potenza*, or the *virtus* of the space of the ban that justifies this ordering and organization of contemporary human communities and their targeting beyond considerations of life and death. Such a virtual power or force is no longer a "power over man insofar as man is a living being,"[77] as Foucault argued (and proponents of a biopolitical perspective accepted at face value). Rather, it is a power that has the "ontological material" of humanity as its principal objective to the extent that it is always already primed for the transformation of "unique beings into a superfluous mass."[78] This superfluousness of the human characteristic of horror and agony (as we defined it in Chapter 1) requires the filtering of human beings through a generalized (because virtual) *nomos* of exception.

This perspective on contemporary power or *potenza* and its violence adds a geopolitical context to the kind of humanity, or perhaps, the kind of emerging life that has to accommodate itself of such a new *nomos*. In Chapter 2, we saw how governmentalized insecurity, or the fear of not living a normalized life, resulted in a form of self-cancellation, a self-cancellation or auto-immunization that prepared the terrain for a humanity to be turned into a superfluous mass. The fear of being afraid helped fuse the body (at the singular and then collective level) with governmental *dispositifs* and gave way to what we called (with Dillon's help) an emergent humanity, a humanity for which individualities are no longer recognizable and that always self-prepares for its horrific dismantling. By highlighting in this chapter the *nomos* of exception that provides a semblance of order/organization to emergent life, and by placing the analysis beyond concrete, material, and bounded biopolitical spaces of exception, we can now understand how emergent humanity is in an important way a virtual humanity too. Indeed, emergent life/humanity is a life/humanity that is potentially readily available to fill the open space of the virtual camp. It is a life that, as virtual, instantiates the presence of a humanity that has always already surrendered to a *potenza* that cannot care if individual bodies survive or perish, as we intimated above. It is also a life (if it still must be called life) that, accommodating itself of and making its own the virtually actualized condition of insecurity, terror, and horror, is only meaningful as that moving target that materializes the "real possibility" of ever more horror, agony, and disfiguration (or what becomes the ultimate "meaning" of the human).

Judith Butler has intimated that global geographies of terror and war take place through a "geopolitical redistribution of corporeal vulnerability."[79] But, in contemporary configurations of a camp that is both nowhere and everywhere, what Cavarero has called "the potential [again, *potenza* in Italian] for a wound [to human bodies] to occur at any time" must also be addressed by critical geopolitical studies (whether they refer themselves to biopolitics or not).[80] This corporeal vulnerability is ubiquitous in geographical redistributions that no longer localize spaces of exception as previous geopolitical orders perhaps sought to do. The ubiquity of corporeal violence in Guantanamo, in multiple CIA "black sites" of torture, in the London tube, or through many more heterotopic sites and scales such as phone lines and communication networks amenable to constant surveillance and digitalized library records available for printing, downloading, hacking, and reproduction point to the need to recognize geopolitical spaces of terror along virtual lines. Moreover, as we saw in Chapter 1, the unbounded heroic violence that characterizes contemporary modalities of violence and agony also operates beyond spatially demarcated and fixed geographies. Indeed, the heroic agonal war machine is never fully bounded in any particular modality or locality of exception, one that is often said to be "over there," in the wild areas of the global war on terror, and "over there" only. But this war machine has a daily modus operandi that relies on constant interventions "in here" as well. In fact, by constantly working on the potential for banning bodies at home as part of the larger global and never-ending war, the war machine's operations, just like the spaces of exception it occupies and proliferates, is both virtually "over there" and "in here" at the same time, in fact, at any time.

The impossibility of a clear cut demarcation in virtual spaces that could tell us when a "war" would be over, where a "war" would take place, and who the enemies in such a "war" would be point to the actualization of a broad nexus between unbounded agonal sovereign warriors, a governmentalized insecurity that operates through an emergent life or an emergent humanity always ready to negate and perhaps maim itself, and a generalized *nomos* of exception that is immanently real even if it is virtually shaped and structured. As we will see in the next chapter, a thorough rethinking of enmity is necessary under such a condition of terror and horror. No longer can the so-called enemy be rooted in an alterity of recognition or rejection. Rather, agonal sovereignty and the condition of dehumanizing horror recognize no enmity and no enemies. Or, to put it differently, humanity itself constitutes the only "really possible" and virtual form that enmity could take in a context in which the pulverization of the flesh of the hapless human multitude is both the target and the objective.

4 The horror of enmity

Rethinking alterity in the age of Global War

> Losing the enemy [according to Schmitt] would simply be the loss of the political itself – and this would be our century's horizon after two world wars. And today, how many examples could be given of disorientation of the political field, where the principal enemy now appears unidentifiable! The invention of the enemy is where the urgency and the anguish are; this invention is what would have to be brought off, in sum, to repoliticize, to put an end to depoliticization. Where the principal enemy, the "structuring" enemy, seems nowhere to be found, where it ceases to be identifiable and thus reliable – that is where the same *phobia* projects a mobile multiplicity of potential, interchangeable, metonymic enemies, in secret alliance with one another: conjuration.
>
> Jacques Derrida[1]

Introduction

Agonal sovereignty often springs forth from the belief that a proliferation of enemies threatens the contemporary polity. Of late, enemies have been presented through various discursive forms as intractable, dangerous, or implacable foes. To confront or eliminate such foes, it is often said that anything short of endless warfare is not only futile, but in fact unacceptable. War efforts must be intense and unremitting since the enemy can appear at any moment and from anywhere (including from within the public sphere). Thus, it is not just the endless character of the Global War on Terror (its relentless quest to eradicate absolute enemies and to end all wars) but also its intensity that renders the actions of today's warriors against evil enemies necessary and heroic (as we saw in Chapter 1). Of course, this heroic dimension also serves to reemphasize the formidability of the enemy and its capacity to be virtually everywhere. The virtual presence of the enemy mirrors the virtuality of the camp dweller, as we theorized in Chapter 3. Crucially, however, this configuration of a formidable (even if virtual) enmity is supplemented by an affirmation that today's enemy is evermore opaque or unspecified, perhaps spectral. It is so in spite of the multiplication of common labels such as "terrorists," "radical evil doers," or "fanatic extremists" that wish to mark or designate the enemy, or at least hope to qualify its absoluteness. Indeed, the virtual manifestation of the enemy reinforces the belief that agonal confrontation must be omnipresent and total. It must be total not just in its absolute destructiveness (allegedly confronting the absolute presence of the enemy), but also in the fact that it must invest all aspects of everyday life, something that proponents of biopolitical perspectives have been keen to emphasize. Thus, the combat of the agonal warrior against this formidable yet elusive enemy must also come equipped with multiple technologies of auto-surveillance and self-monitoring that, as we saw in previous chapters, indiscriminately place individual bodies in banning spaces or exceptional conditions, in virtual camps, that end up actualizing ever new enemies.

Nevertheless, this seemingly "larger than life" (yet, at the same time, very much individualizing) focus on a virtually existential inimical other in the context of operations of agonal war and sovereignty also produces a particular form of alterity. With today's enemy, we are faced with a transformed modality of alterity, one that can never be completely sublimated into the dominant ideological paradigms of Western modernity. While some scholars have attempted to take advantage of the contemporary deployment of this apparently new form of alterity to re-theorize the linkage between the notion of the modern self or subject and the constitution of the political community, there often remains gaps in attempts to theorize the conditions and forms of today's enmity in contemporary politics. One key example can be found in the way Michael Hardt and Antonio Negri have conceptualized enmity in the context of Empire. Gil Anidjar has argued persuasively that there is an absence of conceptual accuracy in Empire, particularly with regards to the place that enmity occupies in what Hardt and Negri theorize to be the borderless global landscape of Empire. As Anidjar writes: "the enemy [in Empire] is both underconceptualized and overconceptualized … [It is both] central and marginal."[2] On the one hand, in the context of global relations of political and economic power/hegemony/sovereignty that constitute Empire, the enemy "is all but disappearing" since it becomes subjected to an "ambiguous banalization" rooted in the perception that war has been transformed into a generalized form of global police action. Yet, the enemy always remains within Empire's gaze, according to Hardt and Negri, since Empire is a modality of power/hegemony/sovereignty that is constantly on the lookout for potential disturbers of the global order it has established. Hardt and Negri's reading reveals that Empire is concerned with deviancy, criminality, or other potential risks of socio-political abnormality that could weaken or fragment such an Empirean order. This theme of enmity derived from the notion of abnormality or counter-conduct is crucial to biopolitical understandings of alterity, as we will see below. By theorizing enmity this way, Anidjar argues that Hardt and Negri display nostalgia for the apparent clarity that characterized previous (supposedly modern) forms of enmity. In particular, Hardt and Negri long for the presence of neatly demarcated boundaries between self and other. And they believe that "the first question of political philosophy today is not if or even why there will be resistance and rebellion, but rather how to determine the enemy against which to rebel."[3] This leads Anidjar to conclude that, "throughout Empire, the enemy is both lamented (we no longer know the enemy, we have mistaken the enemy) and identified, even called for and wished for."[4]

More will be said below about the concept of enmity as it emerges from critical analyses that rely on a biopolitical paradigm. Nevertheless, Anidjar's critical observations about Hardt and Negri's search for a new enemy can be taken as an invitation to re-evaluate and re-engage the notion of enmity and the idea of alterity as they have been mobilized in analyses of a contemporary political condition wedded to a biopolitical framework. Put differently, what sort of enemy or what kinds of virtually inimical and possibly dangerous alterities emerge through biopolitical analytical lenses? And, just as crucially, what modalities of inimical alterity remain unexplored or untheorized as a result of the imposition of biopolitical frames of representability (such as Hardt and Negri's Empire model)?

The first part of this chapter articulates the dominant assumptions that shape the figure of the enemy within what we have called in the "Introduction" biopolitical frames of representability. The Foucaultian approach to biopower and its relationship to violence (put forward most clearly in the last chapter of "*Society Must Be Defended*") rest upon the construction of a racialized alterity (something we started to discuss in Chapter 2). We seek to show the importance attached to this representation of alterity within the critical paradigm that has been called biopolitical sovereignty. To be sure, Foucault's various genealogies uncover the hidden alterities that lie beyond the norm and that, in fact, help to constitute the norm (whether they are alterities found in criminals, delinquents, vagrants, mentally challenged individuals, or racial others). The presentation and production of these alterities give credibility to the implementation and proliferation of biopolitical practices, techniques, and *dispositifs* with a view to governing the tendency towards violence supposedly present in individual bodies or groups. But, as we have seen in previous chapters, attempts to delineate conceptual boundaries, be they between war and politics, normality and fear, or spaces of law and order versus states or conditions of exception, still fail to capture both the multiplicity and the malleability of contemporary forms of agonal violence, terror, and horror. What, then, does the enemy who emerges through configurations of biopower do to our understanding of horrific violence? Is it enough to postulate the presence of absolute enemy bodies and lives or, more generally, of absolute alterities that need to be normalized, regulated, or even banned from the body politic? Is the ability to represent or reconstruct boundaries (and preferably clear-cut ones, as Hardt and Negri, for example, would very much like to see) between self and other, friend and enemy, or normality and abnormality still useful to comprehend contemporary terror as horror?

In the following pages, we examine how the concept of the enemy is operative today beyond the analytical confines of biopolitics and biopower, at the level of a proliferation of allegedly new but, once again, never completely clarified forms of inimical or dangerous alterity, both inside and outside contemporary politics. We explore new formulations and configurations of enemy lives and enemy bodies that are still said to help define the self-identity of an emergent humanity. As we showed in previous chapters, it is quite common for contemporary biopolitical analyses to draw on the thought of political theorists like Carl Schmitt or Giorgio Agamben to situate practices of violence. It is also not unusual for these same studies to reintroduce the notion of the absolute enemy by relying on Schmitt and Agamben. The absolute enmity found in biopolitically inspired texts is often an enmity that cannot be mitigated or brought back into the fold of the body politic. We show that this kind of analytical perspective is particularly dependent upon Schmitt's observation that enmity in late modernity has given way to a proliferation of absolute foes that help to perpetuate a condition of generalized global civil war.

In the last section of this chapter, we argue that the concept of enmity can no longer be rooted in an alterity of either recognition or rejection. Rather, we suggest that agonal sovereignty and its accompanying dehumanizing horror recognize no enmity. To put it differently, it is humanity itself—the fact of being or existing as human—that constitutes the object of any remaining form of enmity at a time when the pulverization of the flesh and the specter of dehumanization beyond any sense of self and other render the search for or construction of enemy lives pointless. Indeed, it is the capacity to exist as human that is the target of horror. And this is, once again, what distinguishes the agonal condition of horror from what often remains a biopolitical context of terror. Whereas terror forces bodies—often enemy bodies—to flee, escape, or be expelled as a result of political or military violence (otherwise, death will await those bodies and lives or, as some have argued, thanatopolitics will replace biopolitics), horror strives indiscriminately to destroy bodies and lives and works to eradicate any trace of humanity. Horror deprives of any humanity both the enemy and the friend, and it ends up fusing them both in an indistinguishable mass of fleshy matter out of which neither recognition nor rejection can be ascertained.

Adriana Cavarero provides us with a hauntingly horrific illustration of such an enmity that both targets humanity and stands beyond it. She relates the story of two 16-year-old girls, Ayat and Rachel, who walked at the same time into a supermarket in Jerusalem in March 2002. Both girls looked very similar, and other people in the store thought they were sisters. But Ayat and Rachel had never met and did not know each other. In fact, despite their physical likeness, they were very much on what many political commentators would consider to be polar opposite sides. Ayat al-Akhras, as it turned out, was a Palestinian suicide bomber who was about to explode herself in the middle of the supermarket. Rachel Levy was a US born Jewish Israeli teenager. Predisposed to be enemy figures, Ayat and Rachel had by chance come together into a place that would eternally keep them together, indifferent and unrecognizable beyond death. Indeed, as Ayat's bomb exploded, only two people died. Ayat herself, and a security guard. Or so it seemed. Rachel, as it turned out, had been killed too, but her body—or, rather, what was left of it—had been so mangled as a result of the explosion that it had fused itself with Ayat's fleshy remains. Only a severed head remained visible, thought to be Ayat's, until Rachel's mother horrifyingly recognized her daughter's face while watching the evening news on television. But the two girls' bodies remained inexorably imbricated, beyond any reasonable identification or rational recognition (except for Rachel's preserved head, but it was first thought to belong to Ayat).[5] Here is a ghastly image of an enmity that did not so much target one or the other, but rather insisted on destroying both the one and the other. Worse yet, it was a form of horrific enmity that made sure that any humanity attached to the body or the life of the one or the other, of the friend or the enemy, would become meaningless, non-retrievable, and forever pulverized as what would be left for all to see were the bits and pieces of the two girls' gruesomely joined together fleshiness. In a generalized context of global agonal war and violence onto the human and human dignity, this chapter provides a critical reflection on what is left of enmity and of enmity's remains and fragments (that, perhaps, are haunted by Ayat and Rachel's own). This leftover of enmity, as a modality of gruesome maiming of humanity itself, is also what critical biopolitical perspectives cannot capture even though they insist on making sense of many instances of contemporary terror and violence onto life and the body.

Recognizing the absolute enemy

The concept of the enemy is a central aspect of Schmitt's notion of the political, as we saw in Chapter 1. Schmitt conceived of the enemy as "the other, the stranger; and it is sufficient for his nature that he is, in a specially intense way, existentially something different and alien, so that in the extreme case conflicts with him are possible."[6] Thus, the enemy could never be an a priori determined singular object. As Mika Ojakangas argues: "the enemy [for Schmitt] is anyone who is concretely disclosed, at the miraculous moment of revealing, as the other and hence, as the real threat to our existence."[7] The definition and recognition of the public enemy is the act of sovereignty par excellence throughout Schmitt's work. This recognition reaffirms the identity of the polity and thereby constitutes the political self (and, by the same token, the concept of the friend). In his *Ex Captivitate Salus*, Schmitt quotes a verse by the German poet Theodor Däubler: "The enemy is our own question as a figure." What this means, according to Schmittian scholar Heinrich Meier, is that

> we know ourselves insofar as we know our enemy and insofar as we define our enemy by defining ourselves. We know him to be our enemy who places us in question or the one whom we place in question insofar

> as we 'know' ourselves, as we make ourselves known to ourselves and to others. The enemy proves to be our friend against his will on the way to self-knowledge, and our self-knowledge is transformed suddenly into a source of enmity when it assumes a visible figure.[8]

The key point in Meier's quotation is the matter of visibility. Indeed, Schmitt stresses the necessity of visibly demarcating the enemy from the friend as a way of making the political intelligible. Schmitt famously argues in *The Concept of the Political* that "the high points of politics are … the moments in which the enemy is, in concrete clarity, recognized as enemy."[9] Again, clear visibility is crucial here.

As we have seen in Chapter 1, we may be able to identify an aporia at the heart of Schmitt's reflections on sovereignty and war and, by extension, on the concept of enmity that emerges in, through, and as war. As Jacques Derrida noted (to repeat what we highlighted in Chapter 1): "[a]s soon as war is possible, it is taking place, Schmitt seems to say … And it is eventual as soon as it is possible … As soon as war is possible-eventual, the enemy is present; he is there, his possibility presently, effectively, supposed and structuring."[10] Without the visible recognition of the enemy by the sovereign, there could not be any notion of a political community made up of a set of friends. Another way of putting it is to suggest that the virtuality of war (or the possibly eventual character of war, as Derrida prefers to say) is essential to the visible and clear emergence of the enemy and to the creation of a political subjectivity (the friend). While, according to Schmitt, the enemy must be "concrete" or a "visible figure" in order to have a presence, the requirement for war to be actual or realized is far less important. Rather, it is war's capacity to be actualized (and eventual) that is the guarantee that both friend and enemy will be materialized and, in fact, will find each other. This intricate conceptual relationship between the virtuality or possible eventuality of war and the actual visibility of the enemy (and of the political) is what political recognition (of both other and self) implies for Schmitt.

Even though such a process of recognition is put in place, Meier asks what happens to the concept of the enemy in Schmitt when/if recognition is lacking or not achieved. As Meier puts it: "Schmitt does not stop for a minute to deal with the enemy who threatens our existence without recognizing us as the enemy or who places us in question existentially without our having to agree to the recognition of his recognition."[11] In other words, what if the enemy does not see us as his/her own enemies but nonetheless is intent on destroying us? What if the friend/enemy encounter and antagonism are not actualized but war still does happen, either as a virtuality or as an actual condition? Put another way, what if the other who is supposed to oppose us is actually not an other to us because, in this case, (s)he refuses us as his/her own enemy? In this situation, Meier suggests, the aporia of war identified by Derrida and Karl Löwith (as we saw towards the end of Chapter 1) gives way to an aporia of enmity in Schmitt's thought, an aporia that potentially postpones the intended link between recognition of the enemy and self-realization (and, by extension, renders Schmitt's essence of the political far more uncertain and far less stable). For Meier, the possibly broken link between recognition of the enemy and self-understanding reveals that, in order to be operative, Schmitt's notion of the political can only approach "the most extreme degree of intensity where it is a question of the battle against an adversary whose moral dignity is contested, whose historical legitimacy is disputed, whose religious orthodoxy is negated, or the battle against an enemy who for his part attacks his opponent as the absolute enemy."[12] Thus, according to Meier, the notion of an absolute enemy is inevitably crucial to Schmitt's analysis. Only the figure of an absolute enemy can resolve the aporia of enemy recognition and self-awareness, as mentioned above. Thus, the absolute enemy is the non-recognizable, non-visible other of Schmitt's political theory, an other who short-circuits the very relationship between enmity and self-awareness that Schmitt seems to want to establish.

Susan Buck-Morss goes one step further and seeks to emphasize the significance of the absolute enemy for Schmitt's concept of the political. She wishes to highlight a Hegelian understanding of enmity based on a postulated "ethical difference" in Schmitt's thought.[13] Buck-Morss argues that the concept of enmity, in Schmitt's *The Concept of the Political* in particular, needs to be revisited in order to recognize the diverse constitution of what she calls "political imaginaries" in contemporary domestic and global politics.[14] By political imaginary Buck-Morss means a "political landscape, a concrete, visual field in which political actors are positioned."[15] Instead of dividing enmity as Schmitt does between a public enemy (*hostis*) and a private enemy (*inimicus*),[16] the concept of enmity is better understood to be based on the opposition between the categories of the normal and the absolute.[17] And Schmitt's own theorization of enemy recognition may in fact exemplify such a distinction between normal and absolute. For Buck-Morss, the normal enemy is an inimical other who is conceivable within a given political imaginary, within a field of reference inside which political actors ultimately judge the validity of their actions and those of others. The normal enemy acts in a way that does not radically alter, threaten, or call into question the essence of the underlying political order. The normal enemy plays by the rules of the political order established according to one's own political imaginary. In fact, this enemy is constituted by the rules of the political order themselves. As Buck-Morss states: "so long as the enemy stays in its place, keeping its position allotted to it within the political imaginary, … it is not a threat in the absolute sense."[18]

By contrast, the absolute enemy is one that radically challenges the ontological foundation of the political community and its imaginary.[19] Thus, not unlike what Schmitt had suggested about enmity, the actions of the absolute enemy can reach the pinnacle of self-insecurity, not simply or even principally in a physical manner, but more importantly in an ontological sense as well to the extent that potential or actual attacks perpetrated by such an enemy "threaten(s) the imaginary landscape itself."[20] For Buck-Morss, the spectacular attacks on 9/11 are one of the clearest examples of what an absolute enemy could look like and could do. On that fateful day, the American political imaginary (about itself and the world) supposedly shifted from one of general ease to one now based on radical insecurity.[21] Those attacks also dramatically revealed the inadequate geopolitical consciousness of a nation-state (and its political imaginary) rooted in the idea of protection and security from an enmity reliant upon clearly demarcated boundaries. But Buck-Morss's own formulation of enmity is not without introducing its own set of paradoxes. One of the key paradoxes of this conceptualization (and of the revelation that the absolute enemy is always already part of political imaginaries that seem only to have room for a "normal" enmity) is that, as Buck-Morss admits, a total negation of enmity becomes in and of itself a threat to "the legitimacy of the sovereign agent."[22] Here, Buck-Morss's theorization rejoins Schmitt's own since an absence or an eradication of the enemy other (normal or absolute) is in fact a danger to the political imaginary itself from whence the idea of enmity emerged in the first place. And since normal enmity could turn into absolute enmity (since Buck-Morss does not provide any clear justification for why this shift could never happen), the possibility of destruction of both the enemy and the political community is always at hand because absolute enmity cannot be reconciled with one's political order (since it is, once again, an ontological threat). Indeed, similar to Schmitt, the essence of the community/political order and the rationalization of the political imaginary or political landscape depend upon the construction of enmity—in fact, an absolute enmity far more than a normal one—because enmity gives the political imaginary its shapes and forms. It is enmity that helps to "position" the political actors into the "visual field." Without absolute enmity (not just regular enmity), the political community runs the risk of dissolving itself.[23] Thus, Buck-Morss appears to go one step further than Schmitt by transmuting Schmitt's "friend–foe" distinction (and its meaning for the constitution of the political self) into an absolute form of enmity. Or, to put it slightly differently, Buck-Morss takes advantage of Schmitt's aporia of enmity to reveal that matters of self and other political construction, identification, or recognition are always about mobilizing absolute enmity.

Thus, the absolute enemy is always a necessity within any geopolitical community or order since, as Schmitt and Buck-Morss indicate (willingly or not), absolute enmity legitimizes the sovereign essence of the political. Absolute enemies function as anchors for a political community because they potentially reactivate the latent sovereign function of deciding not simply on what is inside and what is outside the polity, but also on what must be properly and fundamentally political. Derrida already perceived this crucial relationship between the political and absolute enmity when he argued that Schmitt's concept of the political could not be construed as a self-contained or autonomous concept (despite Schmitt's efforts to present it as such).

Schmitt's positing of the friend–enemy distinction was always already reliant upon an ever fragile, unstable, and possibly dangerous dialectic of recognition. As Derrida wrote:

> But no politics has ever been adequate to its concept. No political event can be correctly described or defined with recourse to these concepts. And this inadequation is not accidental, since politics is essentially a *prâxis*, as Schmitt himself always implies in his ever-so-insistent reliance on the concept of real, present possibility or eventuality in his analysis of the formal structures of the political.[24]

Of course, Schmitt's definition of the political was intended to remain fundamentally structural. The political for Schmitt did not rest upon any sort of psychological or affective conception of friend versus enemy. Instead, as Schmitt stated: "the friend and enemy concepts are to be understood in their concrete and existential sense, not as metaphors or symbols ... least of all in a private individualistic sense as a psychological expression of private emotions or tendencies."[25] Derrida clarified:

> The political enemy would not inevitably be inimical [i.e. the private enemy], he would not necessarily hold me in enmity, nor I him. Moreover, sentiments would play no role; there would neither be passion nor affect in general. Here we have a totally pure experience of the friend-enemy in its political essence, purified of any affect—at least of all personal affect, supposing there could be any other kind. If the enemy is the stranger, the war I would wage on him should remain essentially without hatred, without intrinsic xenophobia. And politics would begin with this purification. With the calculation of the conceptual purification. I can also wage war on my friend, a war in the proper sense of the term, a proper, clear and merciless war. But a war without hatred.[26]

The absence of a conceptual connection between the constitution of the enemy as a structural requirement for the political and any form of psychological or affective condition meant that the concept of the enemy had to have a spectral quality. A purified enemy existing merely as non-friend or as a strictly public or objective category that cannot evoke or be the result of any passionate engagement presents itself as a ghostly figure. Indeed, like a ghost or a specter, Schmitt's public structural enemy haunts the political (and, through its haunting, conditions the decisions made by the sovereign in concrete settings). Yet, it is never to be identified as a really existing antagonist whom the self or friend could intimately recognize because it awakens a series of affective responses for the self (including hatred, as Derrida suggests). The spectral quality of Schmitt's enemy/other already announces the aporetic dimension of Schmitt's political enterprise as it ushers in a political order (or an imaginary, as Buck-Morss would put it) that is defined by a necessity to oppose, go to war against, and ultimately destroy inimical others that exist not so much as objective structural facts but as hauntings.

But the figure of the absolute enemy seems to break down the spectral logic that enables Schmitt's political (and Schmitt's sovereign) to retain a semblance of structural objectivity or necessity. Taking as his cue the thousand years of conflict between European Christendom and Islam that Schmitt mentions in *The Concept of the Political*,[27] Derrida argues that "defending Europe against Islam, here considered as a non-European invader of Europe, is then more than a war among other wars, more than a political war."[28] Because Islam is presented to embody incommensurable differences from Europe, the war to defend Europe against Islam takes on a peculiar dimension, one that is both political (it could simply reflect a friend–enemy distinction rooted in the very possibility of war and violent death) and non-political. Indeed, as Derrida explains:

> this would be not a war but a combat with the political at stake, a struggle for politics … On the political side of this unusual front, the stakes would be saving the political as such, ensuring its survival in the face of another who would no longer even be a political enemy but an enemy of the political—more precisely, a being radically alien to the political as such, supposing at least that, in its purported purity, it is not Europeanized and shares nothing of the tradition of the juridical and the political called European.[29]

Here, the prospect of a non-European enemy introduces another dimension, namely, the possibility that the other may be, as Derrida writes, "radically alien" to the political itself (and not just other to the friend and its political order or imaginary). This reflection comes close to Buck-Morss's own analysis of absolute enmity after 9/11. Indeed, by assuming that this type of enemy actually rejects what counts as the political in a European (Western) tradition even before the encounter with such an enemy can take place (and then can give the political its essence), Schmitt constructs a truly alien figure, a monstrous other that, of necessity, displays elements of private, emotional, passionate, and possibly visceral radical otherness. Such a radical/absolute alterity trumps any sense of friend and foe as a structural political distinction and, in a way, reveals that hatred and a desire to obliterate this absolute inimical other are what haunt Schmitt's politics. Nick Mansfield has reflected on this dimension of Schmitt's theory and he has suggested that the emergence of the absolute other in Schmitt's work makes the concept of the enemy a paradoxical notion. As Mansfield puts it: "this [absolute] enemy … does not only come from outside politics, but is anti-political. They are your enemy and not your enemy, but they are also opposed to the whole concept of enmity, perhaps even the enemy of enmity."[30] Absolute enemies, then, are the enemy of enmity because, as we indicated above, they have already cancelled out, through their very being (actual or virtual) or presence (real or spectral), the possibility that the friend–foe antagonism could serve as the objective foundation for the political.

What Derrida's reading of Schmitt's political and Schmitt's attempt to demarcate pure political enmity from a more radical (non-European) form of otherness points to is the artificiality of a political community founded on an allegedly objective and affectively neutral antagonism that, in order to be preserved as such, must force the specter of absolute enmity out of the political imaginary. But far from being conjured away, the specter/ghost (and its haunting threat to take over the entire political edifice) reveals the fragility, the arbitrariness, and the fabricated origins of the friend–enemy distinction, all the more so since absolute enmity is presented fundamentally to negate or endanger the political community before any friend–foe antagonism has had any chance to define the sovereign political domain. Thus, as the ghost of the absolute other shows us, Schmitt's logic of the political can only work to the extent that it suppresses an original impurity or negation, an impurity or negation that makes us question whether the alleged opposition between enemy and friend exists at all. Indeed, by always already rejecting or cancelling out the political community and its imaginary, the specter of the absolute inimical other undermines the belief that there is or can ever be an inside and an outside, a here and a there, an "us" and a "them," and, moreover, that friends are those who are found in the domain "within" whereas enemies are located in the space "without." Instead, since a political order/imaginary that guarantees a clear demarcation between a space of the friend and a space of the enemy is to be doubted from the very beginning, an indistinction between or even an interpenetration of enmity and friendship is always operative. As a result, and as is far too common with expressions of absolute enmity, today's friend can easily turn into tomorrow's deadly existential other. Or, to evoke a familiar Global War on Terror theme, the enemy is always already walking among us. This last possibility is, according to Derrida, always available as a consequence of the aporia of enmity and, terrifyingly perhaps, often becomes a readily present excuse for xenophobic hostility.[31] As Derrida puts it:

> As in every racism, every ethnocentrism-more precisely, in every one of the nationalism throughout history—a *discourse* on birth and on nature, a *phúsis* of genealogy (more precisely, a discourse and a phantom on the genealogical *phúsis*) regulates, in the final analysis, the movement of each opposition: repulsion and attraction, disagreement and accord, war and peace, hatred and friendship. From within and without.[32]

Whereas Schmitt emphasizes the need to actualize the enemy figure through a dialectic of recognition that serves to construct the political community, Derrida reveals that the aporia of the absolute enemy (with its spectral logic) undermines any attempt at drawing conceptual and political boundaries. Schmitt feared that, without a concrete materialization of enmity, political concepts would "turn into empty and ghostlike [his own word] abstractions when this [concrete] situation disappears."[33] But it is precisely the enemy's spectral haunting that allows enmity to exist, be meaningful, and, if need be, become embodied. What remains undertheorized in many contemporary commentaries that follow Schmitt's path to the political, and his belief on how enmity is generated (and, by extension, on how the idea of absolute enmity creeps in to provoke unbounded violent overreactions aimed at neutralizing this existential other), is this spectral dimension that perhaps emerges more clearly when past geopolitical imaginaries become less secure. What we wish to stress, then, is not the propensity in Schmitt's analyses and in many contemporary commentaries on enmity to actualize or materialize an absolute enemy, an inimical alterity that would legitimize the deployment of agonally sovereign warriors. Rather, following Derrida, what matters is the spectrality of enmity that enables the figure of the inimical other to haunt any enemy materialization. The spectrality of this enemy other, the ghost actually among us (that Derrida further links to the idea of an auto-immunitary violent response), informs many instances of violence and terror that define the Global War on Terror and its concerns with allegedly evil and annihilating enemies in our midst. By developing the linkage between the spectral enemy and biopolitical frames of representability, a matter to which we turn in the next section of this chapter, we can get a better sense of how enmity is not just about identifying and embodying a demarcated and materialized alterity, but also and perhaps more crucially about capturing humanity itself.

Biopolitical representations of alterity

Not surprisingly, most understandings of contemporary enmity/otherness influenced by a biopolitical perspective owe their theoretical postulates to the work of Michel Foucault and, in particular, to Foucault's depiction of the relationship between the normal and the abnormal.[34] Drawing on his early work on madness and the disciplining of bodies, Foucault wrote that "[d]isciplinary normalization consists first of all in positing a model, an optimal model that is constructed in terms of a certain result, and the operation of disciplinary normalization consists in trying to get people, movements, and actions to conform to this model, the normal being precisely that which can conform to this norm, and the abnormal that which is incapable of conforming to the normal."[35] Foucault's insistence on the importance of the norm is not only meant to describe how institutions exclude individuals. Rather, the norm opens up a field of visibility that facilitates the construction of "multiple separations, individualizing distributions, an organization in depth of surveillance and control, and intensification and ramification of power."[36] The norm thus includes as much as it excludes. It includes in an especially intensive fashion by focusing on non-forceful techniques of examination, discipline, and regularized accountability that ceaselessly intervene at the level of docile bodies. When Foucault turns his attention to sexuality and its capture within a modern biopolitics of normalization at the level of an entire population (as he explains in the *History of Sexuality*, Vol. 1), he then argues that the expansion of sexual discourse in the nineteenth century should not only be understood as an attempt to define the moral boundaries of sexual conduct in society. Rather, biopolitical normalization, in this case, acts as a modality of power that intensifies the visibility (and brings into focus) and shapes the expert understanding of various manifestations of sexualized conduct. Thus, as Jeffrey Thomas Nealon puts it: "Foucauldian biopolitical norms do not primarily work to exclude the abnormal; rather, they work ceaselessly to account for it—to render it normal or abnormal—and in addition to link that evaluation with the murky, amorphous category of life or lifestyle." This observation further leads Nealon to conclude that "biopower's norms are efficient and continuous calculations of alterity, not the binary banishment or exclusion of it."[37]

The crucial issue raised by Nealon here is that the rejection of the abnormal (and the corollary intensive capture of the body by regimes of normativity and normality) within biopolitical frames of representation is predicated upon the presence of an epistemological domain that strives to rediscover or reinvent alterities beyond the stated or recognized norm. Alterities are what must be found on the side of the abnormal, an otherness that, from outside the norm, calls in and reinforces the regularity and power of the norm on behalf of life itself or for the sake of the vitality of the population. We have already seen in Chapter 2 how Foucault envisions a biopower that effectively transforms race into racial enmity so that oppositions and even conflicts against such hostile forms of life (though they may not be deemed worthy of living to start with) can take the shape of a total enmity. Foucault makes it clear that the racial/biological enemy is not the product of "old tradition[s], [and is] much deeper than a new ideology." Foucault further argues:

> [w]e are dealing with a mechanism that allows biopower to work. So racism is bound up with the workings of a State that is obliged to use race, the elimination of races and the purification of the race, to exercise its sovereign power.[38]

Foucault's emphasis on race (the constitution and further elimination of "different" or "other" racial categories or the corollary notion of the purification of one's "own" race or species) is emblematic of how biopower constructs, classifies, and relates to enemy life, a life that is also granted another important dimension: that of an absolute enmity characteristic of abnormal, subhuman, or even non-human bodies. According to Michael Dillon and Julian Reid, the idea of an absolute racial/biological enmity is better arrived at through the use of another metaphorical figure, that of the "rogue." Derived from Derrida's theorization of the rogue/roguery (though, it must be said, Derrida does not at all intend to develop a biopolitical argument when he discusses the notion of the rogue),[39] Dillon and Reid's rogue embodies characteristics of a non-correctable subhuman abnormality very reminiscent of Foucault's theorization. Indeed, Dillon and Reid write that "the rogue is marked by its inhumanity, aggression, non-conformity, and disorder."[40] They add that the rogue partakes of a typical biological political imaginary since "the protection of the political order from the threats posed by rogues" amounts to "securing life."[41] For Foucault, and for proponents of a critical biopolitical approach to power and violence who follow Foucault's intellectual lead, it is not uncommon to find the historical example of Nazism prominently displayed as an utmost form of biopolitical roguery or abnormality. Nazism, it is said, represents the "paroxysmal development of the new power mechanism" that turns the rogue or the abnormal into the absolute racial/biological enemy whose life is no longer worth living.[42]

As we hinted at in the "Introduction" to this volume, Roberto Esposito offers a rather similar understanding of a biopolitics of enmity and, through biopolitical enmity as abnormality, discovers the point of transition from biopolitics to thanatopolitics (or a biopolitics of death). For Esposito, it is also the example of Nazism that reveals the convergence between a "therapeutic attitude and the thanatological frame." The Nazis, Esposito continues, "made [*operare*] a deadly incision, in the specifically surgical sense of the expression, in [the German] body."[43] Esposito believes that what governs what he calls the "homicidal temptation of biopolitics" is an intensification of the immunitary response to a threat of "contagion of superior beings by those who are inferior."[44] Here, Esposito (far more than Foucault) reaches the limit point of biopolitical understandings of enmity and alterity. The limit point of this biopolitical representation is the idea that the absolutely abnormal enemy or rogue whose life is not worthy of life presents a threat to the normal, healthy, or vital body because it is a life/body that always already contains death within itself. What makes this abnormal body into an absolute enemy to be destroyed is that, in fact, its life (if it can even be called a life) is not just worthless but indeed deadly (or, more appropriately, death-like). The enemy body is a constant reminder of the danger of contagion and impurity (so crucial to the Nazis) that can weaken and possibly negate the safe, sound, and pure life/body of the self. Thus, biopolitics of necessity becomes a necropolitics vis-à-vis the absolute enemy (an argument that is also central to Mbembe's work) because the immunity of the political self/subject requires that the contagious and impure other be rooted out. For Esposito, then, the thanatopolitical engagement against the enemy (absolute, deadly, and impure as it is depicted to be) is merely an extension of the immunitary paradigm. Like most forms of immunization, the body of the self may not survive the shocking and violent encounter with this deadly or contagious other. But, if it does survive it, the body and life of the self and the community will be stronger and healthier as a result.

What Esposito's immunitary dispositif of enmity implies is an "absolute normativization of life" that, in the end, produces a modality of alterity that is no longer based on recognition or even rejection. Indeed, in the thanatopolitical logic of absolute (thus, endless) enemy confrontation and generalized auto-immunization, the preservation or ultimate salvation of the normal, healthy, or vital body (the body of the self, of the friend) can no longer be assured. Death itself is a way of engaging the absolute other, perhaps the only way that could guarantee both the other's destruction and the immunitary protection of the self or the community. As Esposito puts it: "the only way for an individual or collective organism to save itself definitively from the risk of death is to die."[45] Esposito's exposition of biopolitical alterity/enmity comes close to the reflection on emergent humanity that we developed towards the end of Chapter 2. His analysis represents the endpoint—perhaps the point of no return—of biopolitical perspectives on enmity because, in drawing out the path of the normal and healthy organism's towards the death or destruction of the rogue or enemy other (this is perhaps the vital body's own death drive), it reveals a radical fusion of the lives and bodies of both the self and the other, of "us" and "them," in a context of violent dehumanization that may no longer have much to do with the need to preserve certain bodies or to eradicate absolute others whose lives had already been declared unworthy of being lived.

Put differently, the distinction between self and other at the thanatopolitical limits of a biopolitics of enmity loses its coherence. Indeed, two possibilities remain after such limits are reached. Either the threshold between life and death has been transcended and another non-biopolitical logic sets in. This is, once again, the agonal sovereignty path we have focused on in this study. Or, if a semblance of biopolitical logic must be retained (as scholars like Agamben, Dillon, or Esposito seem to want to do), it must nonetheless be recognized that many of the conceptual foundations of a critical biopolitical analysis have been subverted. The immunization of the self or the community through a fatal engagement with absolute enmity suggests that the racially absolute enemy actually needs to be reinscribed or reinvented in every single body (self or other) in order for regeneration at the level of the species or the race to take place (if indeed it is still a matter of regeneration of life). The problem with this modality of immunization of the body, with this auto-pharmacotic logic of identity (as Larry George may put it),[46] is that the constantly required insertion of the absolute enemy into the self (individual or collective) fatefully confuses and disables any possible categorical distinction. It no longer allows the enemy to be recognized or visible in "concrete clarity." And it blurs the field of representation or visibility of the self too since no structural, psychological, or affective characteristic is ever attributable to the other that would not be already part of the self. Thus, (bio)political demarcations and categorical constructions are rendered meaningless. And, one may suggest, the more pointless this inimical alterity becomes, the more intensively violent yet frustratingly futile attempts to redefine or rediscover it are likely to be.

A spectral enmity?

In a recent essay exploring why the Global War on Terror produced spaces of exception such as Guantanamo, Abu Ghraib, or the many black sites of torture at various US military bases throughout the world, Anne McClintock asked a rather self-evident question: "what is the motive for torturing people whom the government and the interrogators know are innocent?"[47] As she explains, it has become common knowledge by now that a majority of individuals caught up in the web of America's war on terror had no direct involvement with terrorism. Despite this recognition, the practices of apprehension and detention of individual bodies in camps (actual or virtual) continues. McClintock's explanation for why the space of the camp and its continuous inclusion of bodies and lives are not about to end seeks to relate the contemporary politics of exception to the idea of enmity. As she puts it: "[t]he state was faced with an immediate dilemma: how to embody the invisible enemy and be visibly seen to punish it?" She concludes that "the US state had to turn ordinary people into enemy bodies, bodies that could be subordinated to what I call 'super-vision' and put on display for retaliation."[48] This ordinary embodiment of enmity occurred in three ways, according to McClintock. First, the US executive branch needed to give the enemy a "recognizable face," that of Bin Laden to start with and of his cohort of Al-Qaeda terrorists. Second, the Global War on Terror was geopolitically territorialized through US military efforts in Afghanistan initially, and later in Iraq, thus providing a concrete spatiality (and an identity) to many of the bodies that would end up occupying the space of the camp (Afghan Talibans and Iraqi separatists, as well as their global sympathizers). Third, the process of embodiment was designed to produce the "enemy" as a body, with a corporeal appearance, that could then be placed under the constant supervision of the US war machine (primarily), and thus rendered "most dreadfully subject for retaliation and revenge in the labyrinths of torture."[49] Thus, McClintock expresses the view that the mostly US-led war was about reasserting a logic of recognition, delimitation, and rejection onto a specifically corporeal enemy other.

While McClintock insists on the return to the enemy/inimical other as an embodied presence, one that can guarantee recognition (of the self), rejection (of the normalized other), and reconstitution (of the body politic), the immunitary logic of absolute alterity advanced by Esposito asks us to be suspicious of such a return to an (almost Schmittian) enmity, one that would be finally visible in all its concrete clarity. Perhaps instead of insisting on placing a concrete, embodied, and normalized enemy (the Al-Qaeda terrorist, the Iraqi insurgent, the Taliban fighter, the Islamic fundamentalist, the traitor to the US cause, or any other camp dweller) in the context of a US-driven Global War on Terror whose objective mostly would be to rediscover a principle of recognition and reconstitution of the self, we would do well to think about enmity (concrete or absolute) and the thanatopolitical logic of destruction of both self and other in the light of another analytical geopolitical configuration. This alternate conceptual approach is Carlo Galli's notion of "Global War." Global War, for Galli, is not the same thing as the Global War on Terror. Rather, Global War is an analytical space that encourages a general reflection on a political condition in which "there is no longer an institution or boundary that can impede certain types of events from happening on the inside (unpredictable and acute outbursts of violence, for instance), or the loss of security in different domains such as health (threatened by new diseases), access to water, stability of employment, retirement plans, and the economic system."[50] Global War is the projection on a planetary scale of a generalized fear whereby "anything can happen anywhere at any time."[51] In Global War, any attempt at re-embodying the enemy or at re-territorializing conflict is a self-defeating enterprise. As Galli argues:

> Many have tried (and continue to try) to respond to this predicament employing various strategies to attempt to transform Global War into a different kind of war, one that is more comprehensible and more manageable. Some have tried to turn Global War into a war against terrorism that, in turn, would be able to establish a clear difference between "Us" and "Them." However, this very overemphasis on the opposition between "our" identity and "theirs" is in itself a product of Global War, one of its most striking psychological and political effects.[52]

Global War, then, only offers a semblance or a simulacrum of a geopolitical order based upon traditionally distinct friends and enemies. Because in Global War the referentiality of the conflict, combat, or agony (and of the enemy too) is irretrievably lost—the war could equally be waged on the enemy or on oneself, as we saw above—only a desperate or nostalgic attempt at recreating a sense of self and other, of us and them, and of inside and outside can take place (this argument recalls Anidjar's critique of Hardt and Negri's notion of enmity in Empire too, as we saw above). But this attempt is far more the result of wishful thinking, of a will to believe, or even a way of covering a psychological loss than it is the product of actual conditions of war antagonism.

Consequently, Galli suggests that the condition of Global War introduces a new concept of enmity. In a manner reminiscent of Derrida's reading of Schmitt's notion of the enemy, Galli turns to the image of the "phantom-enemy" to explain what otherness or alterity look like in Global War. Galli's phantom enemy is an absolute enemy. But here it is absolute because it reveals itself as a specter of absolute indistinction between friend and enemy, between self and other. In a way, Galli's phantom-enemy is the haunting specter that gives rise to Schmitt's fear of an undecidable condition whereby it would no longer be possible to distinguish concretely between a friend and public foe and, as a result of such a nightmare, the political would remain without foundation. As Galli observes: "the obsessive and vain search for security pushes to mark borders that, in turn, with their excluding and hostile arbitrariness, end up demonstrating that even traditional identities become phantoms in the globalized space."[53] Invoking a language and imagery of ghostliness and spectrality, Galli goes on to state that

> The more we want to reassure ourselves about these specters, the more uncanny they become. The phantom-enemy is so irrepresentable that even its hyper-representation is difficult: it is not by chance that the opposed propagandas are more preoccupied with the virtual construction of the friend's identity than with the enemy's representation. The phantom-enemy is the ultimate figure of hostility and the most uncanny because it is not different or distant from us. Rather, it is our shadow, our anguish, the dark side of our own identity—itself made uncertain and phantasmatic by the gray global twilight.[54]

Galli's insistence on the spectral dimension of the contemporary enemy is not without recalling (as specters often do) what Gilles Deleuze and Felix Guattari in *A Thousand Plateaus* referred to as the "unspecified enemy." According to Deleuze and Guattari, the "unspecified enemy" is also a generalized condition of alterity. Moreover, it is a condition of alterity that characterizes what Deleuze and Guattari would later call "control societies," that is to say, societies that stand beyond biopolitical imperatives.[55] Deleuze and Guattari's "unspecified enemy" is beyond biopolitics because it has no body and no life. And, like Galli's spectral enemy, the unspecified enemy has no death either. Such a figurative, virtual, yet very present and insistent inimical other (present and insistent because contemporary wars are being fought both against it or on its behalf) dwells beyond life and death, beyond embodiment and spiritual transcendence too. Indeed, the phantom/unspecified enemy is a spirit and a presence, but a ghostly spiritual presence or condition of (non)being. Zombie and undead like, the phantom/unspecified enemy incanted by both Galli and Deleuze and Guattari is also a certain figure of humanity and the human that, beyond humanity itself, can entertain the idea of the non-human or "anti-human" (to use Primo Levi's turn of phrase). Thus, the spectral/unspecified enemy possesses what Brian Massumi considers to be a quasi-infinite form that an enemy as other (or self) can take. As Massumi argues, this enemy is "humanly ungraspable."[56] It is, Massumi adds, a "faceless, unseen, and unseeable" enemy that "operate(s) on an inhuman scale."[57] The idea of such an ungraspable or even inhuman inimical other that could also very well be me (as a result of the general principle of indistinction between self and other that governs it) is a key insight, because it allows us to raise a series of new questions about the relationship between the self, the state, and the violence perpetrated onto others.

Massumi notes that "[t]he enemy question is not who, where, when, or even what. The enemy is a what not—an unspecifiable may-come-to-pass, in another dimension. In a word, the enemy is the virtual."[58] Questions regarding who or what the enemy is, or even if it is normal or absolute, miss the target. They do so because they do not speak to the spectrality of the enemy that haunts contemporary (geo)political enterprises. Those questions are also off the mark because, to repeat, they insist on discovering and often reconstructing enemy lives and bodies from which our own biopolitical existence (as a race, a species, or a body politic) could be protected. Beyond the biopolitical or even thanatopolitical confines of enmity, the "what not" question is indeed, as Massumi indicates, what needs to be asked. What the enemy is not yet, or, better still, what the enemy may come to be is what we would like to call an inhumanity-to-be (à venir). This inhumanity-to-be does not just antagonize the idea and figure of the human because, in fact, it may well be what is left or what becomes of the human. Rather, it rejects or, perhaps, abjects any sense of a humanity that requires to be embodied in and concretized as a self versus other or as a friend versus enemy mode of recognition. Indeed, it is the notion of a humanity whose respect for human diversity and individual dignity relies upon established differentiations between selves and others and always already comes equipped with a sense of which attributes are more likely to maintain this idea of a diverse/plural humanity that enmity as inhumanity-to-be challenges and, sometimes, seeks to violently obliterate.

To counteract the principle of humanity as the mark of the one, the individual, or the different, of humanity as the cradle for the *ontos* (as we put it in the "Introduction" to this book) or as that which enshrines the figural unity of the one singular human body and life, the becoming inhuman of contemporary spectral enmity gruesomely displays how indeed humanity can be disfigured and refigured as an undifferentiated mass of fleshy matter. This is how this inhumanity-to-be has recourse to horrific violence and to the dismantling of bodies and lives that could somehow still be recognized or identified as unique, singular, specific, different, or distinguishable from others. Put succinctly, it is to horror (as, once again, Adriana Cavarero theorizes it), to both horror's visual regime and to its paralyzing yet haunting pulverization of the human flesh, that this enmity as inhumanity-to-be turns. A dehumanized and disembodied flesh that could be human, animal, or both, that could belong to any dead or living thing, becomes the unbearably horrifying mark of an enmity that has the human as its chosen target. What took place in the concentration camps in the Second World War could be taken as an obvious illustration (both in its mode of operation and in its horrendous effects) of this "what not" of enmity, of this ghostly, spectral, and unspecified enmity.

In the context of today's Global War, actions by the mostly American war machine and some of its agonal warriors have also evoked such an enmity as inhumanity-to-be. Recent reports have brought up the ghastly story of a few US soldiers deployed in Afghanistan (and, for some of them, in Iraq before) who were sometimes nicknamed the "bone collectors" or the "kill team."[59] Those soldiers (a dozen of them) from the Army's 5th Brigade, 2nd Infantry Division at the Lewis-McChord base in the State of Washington have been put on trial for war crimes they may have committed while on duty in Afghanistan. There, under the orders of Corporal Jeremy Morlock and Staff Sergeant Calvin Gibbs, these soldiers apparently randomly picked Afghan civilians to kill them (later trying to cover the kill by claiming that the civilians were Taliban sympathizers). But the kill was not enough to Morlock and Gibbs. After killing their victims (how many victims overall remains to be determined; Gibbs is also under suspicion for having carried out similar operations in a prior tour of duty in Iraq), the troops often chose to dismember and mutilate their preys, and to scatter their remains. Moreover, several of the soldiers decided to collect a few of those human artifacts, mostly bones (finger bones, femurs, skulls), that they sometimes brought back with them to the United States.[60] More than war crimes, these actions (if proven to be true, and there is ample evidence that these acts did take place) are a form of crime against humanity. Though they do not fit what international criminal law has in mind by the notion of a crime against humanity (a systematic plan to annihilate a population on a massive scale), they could nonetheless be considered actions that reflect an enmity that targets humanity or the singularity of the human. These horrific acts are expressions, in our account, of an enmity as inhumanity-to-be for whom the death of the other is not sufficient because to randomly kill others would be, to return to Cavarero's words, "too little" and would not aim at the "uniqueness of the body."[61]

The fact that such an enmity as anti-humanity also evokes the motif of the hunt (hunting down the human, trapping the human prey, cutting off and dividing the human catch among the hunters, or collecting and displaying the trophies of the human hunt) is perhaps not accidental either. In his recently published (in French) critical genealogy of the hunt, philosopher Grégoire Chamayou suggests that the logic of the hunt (for any prey) reveals a predatory power.[62] Such a predatory power, or what Chamayou calls a "cynegetic" force or a modality of "cynegetic sovereignty,"[63] can be opposed to the pastoral power (introduced by Foucault in his lecture *Security, Territory, Population*) that insists on protecting or shepherding life at all costs and can be seen as a foundation for biopower.[64] By contrast, "cynegetic" power, the power of the hunt, is a power of capture, a force of immobility.[65] While the hunt displays movement in its exercise (and, in this phase, it terrorizes the prey), the hunt also seeks to render the prey immobile, horrified, seized, and killed. "Cynegetic" power seeks to stop motion and wants to put an end to the flow of bodies in space. Indeed, according to Chamayou, the "cynegetic" power revealed in the hunt seeks to accumulate bodies in one place,[66] to pile them up and amass them on top of one another (this amassing modality of the hunt may recall the image of the "human pyramids" made up of the Iraqi captives' bodies displayed in the infamous Abu Ghraib photos). Chamayou goes on to suggest that "cynegetic" power is limitless because it expresses a logic of excess and expenditure.[67] Not only is the hunter never satisfied by the one kill, but the hunter always finds more bits and pieces to pick up, gather, and collect from the bodies it has brought down.

While Chamayou still believes that the hunter goes on collecting and consuming "until exhaustion and until death,"[68] it is in fact much beyond death (beyond the death of the prey, or possibly beyond the death of the hunter too) that such a "cynegetic" power extends. As such, the "cynegetic" power of the hunter or, better yet, of the hunt itself, is akin to the enmity as inhumanity-to-be that we theorized above. Similar to the spectral enmity that increasingly characterizes the many instances of violence and antagonism in Global War, the hunt (as Chamayou describes it) targets and captures the human in order to dismantle its figural unity. "Cynegetic" power, just like the spectral enmity that announces an inhumanity-to-come, seeks to "tear at the constitutive vulnerability" of the one individual body or life and,[69] in its place, tries to establish the logic of "the scattered" (*l'épars*), as Chamayou puts it.[70]

Conclusion

Similar to Agamben's claim that "we are all virtually *homines sacri*" (as we discussed in Chapter 3), we could now say that "we are all virtually spectral enemies." Advancing the argument that we are all, potentially, virtually, "really possibly," spectral enemies indicates that the alterity of today's enemy encompasses and, in fact, targets humanity in its entirety. Spectral, the enemy body no longer exists or matters. Or, if it does, it is only virtually, as a representative of the "real possibility" of agony, devastation, dehumanization, and horror. Put differently, today's enemy has no face, no location, no identity. More than absolute or total, this enemy is indeed a phantom or a specter. To return to Galli's useful conceptualization, in Global War, the phantom enemy is a humanity that opens itself up to its own inhumanity, to its own ontological undoing. Under conditions of Global War, enmity and humanity are virtually undistinguishable, and they offer themselves (in an active modality of emergent human life that, as we argued at the end of Chapter 2, is always available for its own dehumanization) as objects of agony and horrific violence. Or, perhaps, neither completely a subject nor fully an object of horror and agony, enmity as inhumanity-to-be enters the domain of the abject that, in their own distinct ways, both Adriana Cavarero and Julia Kristeva have sought to bring into focus.

As Kristeva famously argued, the abject "is not an ob-ject facing me, which I name or imagine. Nor is it an ob-jest, an otherness ceaselessly fleeing in a systematic quest for desire."[71] She added that "[t]he abject has only one quality of the object—that of being opposed to I."[72] But such an opposition does not enable antagonism or duality. Indeed, as Kristeva further specified, the abject also conveys a "weight of meaninglessness about which there is nothing insignificant, and which crushes me. On the edge of non-existence … [the abject] annihilates me."[73] Today's haunting enmity as an inhumanity-to-be is also a becoming abject, a becoming crushed and annihilated, or a horrific abjection-to-come. This horror-to-come is one that, fatally perhaps, presents itself as an abject response to ongoing (and increasingly popular) calls for a democracy-to-come in critical theory circles. For today's humanity and its future, the becoming abject of virtually every human's enmity is the accursed share of the *à venir* (to-come) of any possible democratic event, of the event of democracy, as it has been advanced of late (and perhaps called for) by philosophers like Jacques Derrida or Jean-Luc Nancy.[74] Such a becoming abject of enmity is on the lookout for humanity, for the singularity of the human, as we stated above, but also, along the way, for the possibility that humanity may present itself as the foundation for plural communal living and being (in other words, democracy). In this way, the becoming abject of enmity that informs many instances of agonal violence, horror, and sovereignty is not just ontological (a negative ontology, to be sure, as we suggested in the "Introduction" to the volume). It is also political, or better yet, ontopolitical (or about the violent "construction and sustaining of a political identity,"[75] even if, in this case, it is likely to be a non-identity, or an anti-identity even).

To qualify such an abject enmity as an inhumanity-to-be or a horror-to-come as absolute (as theorists, like Schmitt or Buck-Morss, eager to maintain a distinction between a normal inimical otherness and an absolute or total enmity do) is beyond the point. Further, to suggest that such a spectral enmity seeks to dislocate any sense of a safely recognizable enmity ends up being a tautological claim. Beyond seeking out targeted bodies and lives that, in the process of the search (or is it the hunt?), obtain the label of enemy bodies or enemy lives (perhaps in the hope that such a labeling exercise will facilitate the protection of one's privileged way of life), and beyond insisting that a generalized absolute enmity serve a political function (again, tied to either life or death) in the so-called contemporary biopolitics of terror, the spectral enemy and its abjection-to-be force us to realize that the ontopolitical privilege of the human emphasized throughout Western modernity (even when such a modernity is revisited as a biopolitical modernity) can no longer be assumed or taken for granted. As the illustration of the "bone collecting" soldiers above showed us, the human no longer holds a superior or privileged position over the animal or the bestial in certain instances of horrific enmity/inhumanity. Both the human and the animal, in their indifferentiation, become valued as fleshy matter and, in fact, do "matter" because of their potential as manipulated, dissected, and amassed flesh. Such a pulverized flesh is no longer that of the multitude that Hardt and Negri sought (somewhat naively and melancholically) to reify and revive in their work.[76] Nor can such a fleshy matter of indistinguishable organic origins count as a human flesh that could become the ferment for the hopeful rebirth of a new humanity (as we often encounter it in organic theories and theologies—from the Christian bible to Hannah Arendt, for example—that construct analogies between the idea of genesis or natality and new social and political beginnings). Instead, this flesh that now "matters" as evidence of the undoing of humanity is the pulp, the sedimentation, or the released waste of all those things human that have been crushed in Global War, often by an agonally sovereign war machine that only answers the calls of an abjection-to-be and a horror-to-come.

In his essay "Politics as a Form of Expenditure," Achille Mbembe argues that what he calls "a new form of governmentality" is presently at work on the African continent (in particular). This peculiar, brutal, and ruthlessly destructive "governmentality" has the massive proliferation of massacres of innocent populations as its central objective.[77] This new form of governmentality takes the flesh of the multitude as its abject target too. It does so, according to Mbembe, to manage "the mobility of … multitudes, largely by extrastate jurisdictions or armed formation."[78] Among the defining marks of this "governmental management" of the flesh of the multitude are techniques such as the forced transfer of populations across geographical spaces and the arbitrary erection of zones of internment and encampment that, often unexpectedly, immobilize bodies and facilitate corporeal mutilations, rapes, and large scale massacres. Mbembe argues that this horrendous violence onto the African body follows "the model of old human sacrifices."[79] Mbembe's invocation of sacrifices is particularly pertinent.

Human sacrifice is indeed a key dimension of this spectral enmity-to-come. Mbembe is probably not correct to assume that this abject horror-to-come that targets the human and seeks its complete sacrifice is another (perhaps a new) modality of governmentality (just like it is not enough to argue, as he does, that thanatopolitics or necropolitics is merely the flip side of biopolitics). "Managing populations" in the way described by Mbembe has much more in common with practices of agonal war-making and warrior horror than it does with more conventional theorizations of governmentality (particularly along the line of the pastoral care of populations as Foucault originally introduced the idea of governmental power).[80] The original element in Mbembe's formulation of this so-called new (perhaps "African") governmentality is that it recognizes the logic of Global War (advanced by Galli, as we saw above) and it assumes the notion of a spectral enmity-to-be that takes humanity as its object (or, better yet, as its abject target). Put differently, in this essay, Mbembe is led to rethink the relationship between absolute or total war violence and the possibility of ever being able to reach realizable ends or objectives. This is precisely why Mbembe chooses to turn to George Bataille's notion of sacrificial "expenditure" instead.

For Bataille, sacrificial expenditure or excess is a by-product of how energy organizes and dominates human life. As Bataille explains: "[t]he living organism, in a situation determined by the play of energy on the surface of the globe, ordinarily receives more energy than is necessary for maintaining life … if the system can no longer grow, or if the excess cannot be completely absorbed in its growth … it must be spent, willingly or not, gloriously or catastrophically."[81] According to Bataille, the notions of expenditure and excess render the anthropological perspectives on human sacrifice (such as those introduced by René Girard, for example[82]) purposeless. Indeed, in Bataille's sacrificial logic (unlike what can be found in many anthropological approaches to sacrifice as a communal ritual), there is no longer any community or any "healthy body" to save or protect, or even to rescue from extreme evil. Rather, sacrificial expenditure for Bataille evokes a proliferation of various (sometimes extreme) forms of violence against the flesh that operate according to a logic of circulation of energetic surplus or excess. Such a violent excess is no longer, if it ever really was, tied to any achievable, let alone rational, ends. Although Mbembe still wishes to articulate violence within a framework of embodied enmity or, as he puts it, of "the idea of an enemy, a 'foreign body,' that must be expelled or eradicated," his theoretical overture towards Bataille brings him closer to our perspective on an abject enmity as inhumanity-to-be. This perspective fatefully removes any need to see the alterity of enmity as something foreign, absolute, or to be recognized. Indeed, in today's context of agonal sovereignty and agonal war, the Bataillian notion of a sacrificial inhumanity evokes the "real possibility" of a mutilated flesh whose horror can only be expanded and be made ever more excessive.[83]

Conclusion

Facing horrific violence

> What is a supplice? "Corporeal punishment, painful to a more or less horrible degree," said Jaucourt in his Encyclopédie article … " It is an inexplicable phenomenon that the extension of man's imagination creates out of the barbarous and the cruel." Inexplicable, perhaps, but certainly neither irregular nor primitive.
>
> Michel Foucault[1]

The grostesque horror of the human supplice

Michel Foucault opens his master opus *Discipline and Punish* with what is perhaps the most famous horrendous depiction of corporeal punishment performed onto a human body. Narrating the *supplice* of the regicide Damiens on the Place de Grève in Paris in 1757, Foucault details the public spectacle of horror that was Damiens' punishment, torture, and slow execution both until and after his death.[2] The ritualized ceremony of unspeakable brutality that befell Damiens had been planned ahead of time by a penal system that sought to target (and take revenge on) the individualized human body. Damiens was to be "conveyed in a cart, wearing nothing but a shirt, holding a torch of burning wax weighing two pounds'."[3] Placed on a scaffold in the Place de Grève, Damiens' flesh would first "be torn from his breasts, arms, thighs, and calves with red-hot pincers, his right hand holding the knife with which he committed the said parricide, burnt with sulphur, and, on those places where the flesh [were to] be torn away, poured molten lead, boiling oil, burning resin, wax and sulphur melted together."[4] Finally, the condemned's body would be "drawn and quartered by four horses and his limbs and body [would eventually be] consumed by fire, reduced to ashes, and his ashes thrown to the wind."[5] Foucault recounts that the horses did not manage to pull Damiens' limbs apart after repeated attempts and that, as a result, one of the executioners had to pull out his own knife to "cut the body at the thighs instead of severing the legs at the joints" as had been planned initially.[6] Deprived of his limbs, Damiens was then thrown on a burning pyre until his flesh was finally reduced to ashes. According to one of the executioners,

Damiens was still alive as he was thrown onto the pyre. Officers were to guard the burning flesh until it had completely turned into ashes and been blown by the wind.

The public horror of Damiens' *amende honorable* (honorable amends) was the product, as Foucault explains, of a justice system that, until about the end of the eighteenth century in France, understood the body to be the "major target of penal repression."[7] In the spectacle of the *supplice* and the horrific punishment of the criminal, it was the power of the law and justice that had to be revealed, a just albeit retribution-driven power that paralleled the centralizing force and authority of the state, sovereign, or government. The horror of the public punishment of the condemned was meant to recall the omnipresence of a legal and political power that could reduce the body to nothingness in the name of justice.

Two and a half centuries later, the spectacle of horror and the *supplice* of "condemned bodies" resurface in Mexico, right along the border with the United States. There, in the ongoing war between drug cartels and cartel-related gangs (or *narcos*), the hideous savagery of human executions, tortures, and destruction of the flesh beyond death is spreading from the California end of the US–Mexico border all the way to the Gulf of Mexico. In what one chronicler has referred to as scenes of "human sacrifice,"[8] cartel and gang members terrorize each other and the local population by way of dismemberments of human bodies, whose bits and pieces are later scattered throughout the border landscape for all to see. In this landscape of violence, death, and horror, the grotesque and the banal come together in a way that may recall the at once absurd and commonplace visual excess of the *supplice* in the French eighteenth century penal system. Among the violent cartel groups and gangs in Mexico these days are Los Zetas, a band of renegade soldiers/warriors who originally acted as enforcers for the powerful Gulf Cartel but lately have split from their former "parent" organization. Responsible for some of the most gruesomely inventive ways of targeting human bodies and the human flesh, the Zetas are also known for their ability to publicize their deeds through various visual media (You Tube videos, in particular). Journalist Ed Vulliamy recounts one episode (among many) of Los Zetas' terror and horror: "In Ciudad Juárez, in November 2008, as dawn broke over the desert, a body was found hanging from a highway overpass. It had been decapitated and was dangling by a rope tied around the armpits. It was still there an hour later, when I saw it—swaying in the wind, hands cuffed behind its back."[9] Other *narcos* groups in Mexico have been known for years for their grotesque use of violence. For example, a Western Mexico cartel referred to as La Familia

> made its "coming out" known in a famous episode: bowling five severed heads across the floor of a discotheque. In Sonora in 2009, a white S.U.V. was found abandoned, and inside it a butcher's display of mutilated bodies—hacked, chopped, castrated, decapitated. It was a carful of human

> cutlets, with no apparent relationship of one piece to another until they were matched by forensic authorities. Earlier this year, 36-year-old Hugo Hernandez was abducted in Sonora; his body turned up a week later in Los Mochis, Sinaloa, but not in a single piece. His torso was in one location, his severed arms and legs (boxed) in another. The face had been cut off. It was found near city hall, sewn to a soccer ball.[10]

Los Zetas, however, have contributed to an intensification of horror that can no longer be linked to what was often thought to be the main rationale behind the cartels' violence: to terrorize the populations and rival gangs in order to exert or maintain control over the illicit drug routes from Northern Mexico to the United States. For Los Zetas, the spectacle of horror, and particularly the scenes of dismemberment, decapitation, or burning and butchering of the human flesh, seems to have no purpose. As Vulliamy writes, it is a horror that is almost "recreational" (hence, the amusement of displaying it online for all to see), performed onto bodies "mainly for its own sake." This makes the drug war along the US–Mexico border, Vulliamy concludes, a war that "is, in the end, about nothing."[11] Or, better yet, it seems to be about nothing but horror itself.

Damiens' public *supplice* on the Place de Grève and the highly visible disfigurements of human bodies by Mexican *narcos* groups like Los Zetas or La Familia seem to have a lot in common. To start with, both rely heavily on regimes of visibility and spectacularity that "shock and awe" witnessing audiences. The visible economy of these two sceneries of violence is not just about terror (to force bodies to flee or to escape a terrifying sight) as opposed to horror (to prevent the body from running away through the paralyzing effect of the gaze confronted with a horrific event), as Adriana Cavarero had initially intended to distinguish between these two modalities of violence.[12] Rather, in a way that recalls Grégoire Chamayou's description of the power of the hunt (introduced in Chapter 4),[13] both Damiens' execution and the cartels' brutal killings and butcherings combine elements of terror and horror. Like the cynegetic force of the hunt, they scare away but also seek to recapture and seize their preys (including those who are expected to watch in horror). They mobilize terror to display the futility of running away and to remind their human victims (the bodies and lives that eventually, too exhausted perhaps, surrender to the violence) that their humanity has already been captured and rendered powerless.

The two "arsenals of horror" (Foucault's own turn of phrase[14]) described above share another common dimension. Both mobilize a violence that is in excess of what at first may have appeared to be their intended objective. Not only is excessive force or excessive brutality obviously at play in both settings. But, more crucially, a violence without limits, ever inventive of new gruesome ways of killing or destroying the flesh beyond the threshold of life and death, is blatantly on display in both Damiens' *supplice* and the cartels' theater of human sacrifice. Here, again, we encounter a theme developed in prior chapters: that of a horrific violence in excess not just of its means and methods, but also of the life and death of the human or of humanity itself. Indeed, the early modern and late modern *supplices* are not only a matter of over-killing, of brutalizing the body over and over beyond the mere biological fact of death (burning Damiens' dead flesh and making sure it is reduced to ashes; the Zetas or La Familia carving out bits and pieces of dead bodies to scatter them about). They are about an excess of humanity, about a wish to exterminate the singularity of human experience and existence. They are about the condition of horror in Global War (as Galli understands it), when wars and conflicts have become "post-political,"[15] and the violence they display have "no single propelling cause, no single objective, and certainly no grand ideology" (as Vulliamy characterizes the so-called drug wars that provide the background for Los Zetas' horror).[16] When violence onto the body is no longer political or ideological, only humanity remains as that which can be destroyed, maimed, or pulverized.

This last observation is perhaps why, despite the apparent resemblance between the pre-modern and the late-modern *supplice*, Foucault's narration of Damiens' grotesque death and the contemporary violent exactions of Mexican drug cartels and paramilitary groups also represent different logics. Indeed, despite the unbearable horror of Damiens' human sacrifice, the ceremonial of his dismantling as an individual body is still supposed to serve a larger, mostly political, purpose. Deprived of any shred (literally) of humanity, Damiens' *supplice* affirms the presence of an all powerful justice system and sovereign power (vindictive as this power may be) that still attempts to rule or govern by means of awesome force and by visual deterrence. This is why, for Foucault, the *supplice* is not "irregular" even if it is still somehow "inexplicable" (as Foucault's quotation at the opening of this "Conclusion" indicates).[17] The early modern *supplice* that targets the condemned's body prior to and after his painful death is still, according to Foucault, a technique of power. It is a crude and dangerous technique, to be sure, even for the sovereign who may not be able to maintain the desired "awe effect" on the terrorized audiences (Foucault mentions that part of the reason why the spectacle of the punishment was abandoned by the French penal system is that crowds had a tendency to sympathize with the tortured victim). But it is a technique that measures and calculates the amount of pain to be inflicted onto the body in relation to the severity of the perpetrated crime.[18] Thus, despite the fact that Foucault states that the *supplice* is a removal of humanity, he also insists on the fact that, in its exercise, the *supplice* never breaks the life–death threshold. As he puts it: "death-torture is the art of maintaining life in pain."[19]

For Foucault, then, the biopolitical motif remains crucial to the spectacle of horrific violence applied onto the condemned's body. It is life that, for him, is the main objective of the *supplice*. And the human sacrifice of Damiens and other criminals like him reflect the existence of a regime of legal and penal power that not only "manages" life to the point of death, but also "subdivides" the life of pain of the criminal "into a 'thousand deaths' by achieving before life ceases 'the most exquisite agonies'."[20] This stubborn emphasis on Foucault's part on the matter of life (biological life, or a life reduced to a condition of bodily pain or its absence), even in the face of the most gruesome, indeed inhuman, violence applied onto both life and the singular body, possibly previews his allegiance later on to a biopolitical frame of understanding. It may also explain why Foucault appears to be far more interested in highlighting the visually public or spectacular part of the *supplice* than its horror (Foucault refers to the "punishment-as-spectacle" as a "confused horror"),[21] a publicity that allows him to retain the focus on the fact of life and on the body.

At the same time though, Foucault appears to be uneasy or at least at a loss for words when dealing with the continuation of the *supplice*'s horror beyond the death of the condemned, once the body is reduced to shredded or burnt flesh. This unease, we would venture to suggest, is indicative of Foucault's inability to conceptualize horror or a certain form of violence that, once again, does not limit itself to the body or life in pain. Foucault does recognize "the very excess of violence" (his phrase) that characterizes the horror of the *supplice*.[22] But he also remains satisfied with claiming that such a violent excess that forces "the guilty man" to "moan and cry out under the blows" is the mark of a "justice being expressed in all its force."[23] Still, the butchered flesh no longer moans or cries out. Yet, this butchered flesh, as the repulsive actions of the Mexican drug cartels repeatedly show us, is also the object of an "excess of violence," perhaps primarily so. Foucault is left to surmise that "those tortures that take place even after death: corpses burnt, ashes thrown to the winds, bodies dragged on hurdles and exhibited at the roadside" are signs that an all powerful (bio)political justice system seeks to "pursue the body beyond all possible pain."[24] But there is hardly any body or any life to pursue, capture, and torture at this point. If nothing else, there is a scattered body, or better yet, scattered bodily remains that may or may not be recognizable as human anymore, and hardly serve any political system or regime of power. Put differently, Foucault's pre-biopolitical (he has not yet discovered the notion of biopower by the time of the writing of *Discipline and Punish*) analysis of horrific violence performed onto the body of the condemned both before and after his/her biological death lacks the necessary language or vocabulary (perhaps not the imagery) that could allow him to make sense of the horror and inhumanity of the scattered (or of the logic of *l'épars*, as Chamayou puts it[25]). Perhaps the late modern or post-political theater of human sacrifice provided by cartels and gangs like Los Zetas and La Familia in a more contemporary context can take the analysis of the *supplice* initiated by Foucault towards the kind of conceptual terrains that Foucault-influenced biopolitical perspectives and frames did not manage to get to.

Indeed, as intimated above, the horror perpetrated and made visible by the *narcos* along the US–Mexico border claims no higher purpose than to spread terror and keep populations paralyzed by fear and incomprehension. Again, initially intended to bear the marks of rival cartels' paths to leadership and control over the border area and the drug routes, the *supplice* of dozens of human victims on a daily basis and the display of their postmortem agony have become meaningless. At best, the deaths, destructions, and dismemberments appear to be "recreational," as Vulliamy indicated.[26] At worst, they reveal a disdain for not just the biological fact of life (such a disdain could be satisfied by putting to death), but also for the very idea of the human, for the belief that there is or has to be something special or unique about humanity. Objectifying the human as they would any other living or non-living thing destined to be used, abused, and discarded, groups like Los Zetas are in total control of their mobilization of violently gruesome means of obliteration of anything that crosses their paths. In this way, they *are* agonal warriors who are left completely free (until they too are pulverized and reduced to bits of fleshy matter or ashes) to transform the border landscape into a landscape of terror and horror.

With bands of armed and brutal individuals like Los Zetas or La Familia, we indeed rediscover the theme of agonal sovereignty introduced earlier in this volume. The *narcos*' agonally sovereign performance is not beholden to any traditional or even non-conventional source of power or authority (certainly not the state, but not even former "parent" cartel organizations and hierarchies). Rather, in a way that is reminiscent of Gilles Deleuze and Félix Guattari's war machine,[27] groups like Los Zetas carve out their own sovereign territory, establish their own authority, and display their own "post-political" performance as they go along, on a day-to-day basis, through every new killing, butchering, or scene of human destruction.[28] The sovereign performance of those agonal warriors (and heroes too for some, including themselves) thrives on the scattering of body parts, life fragments, or memories of human experience/existence reduced to horrifying traumas. As agonally sovereign warriors displaying a sovereign majesty that is as pathetically ghastly as it is heroically ephemeral, drug and paramilitary groups of horror-driven men and women (and sometimes children too) have found ways of generalizing the spectacle of the *supplice* into "abyssal acts of violence" (as Slavoj Žižek put it;[29] see Chapter 1) that periodically reenact the rituals of an inhumanity-to-be, as we intimated at the end of Chapter 4. Just like the ghost-like or spectral enemies that propel it (indeed, who are the enemies of Los Zetas? Are Los Zetas not doomed to turn into their own enemies?), this inhumanity-to-be of the agonal warrior is without limits and bounds. The abyss of horror is bottomless. Only the pulverization of one of the warriors' own flesh can put a temporary end—more of a pause, really—to the horrifying march or hunt of the warrior, to the grotesque scenes of *supplices* and disfigurements, and to the agony of violence, until another warrior, another war machine, another self-proclaimed hero and self-serving agonal sovereign starts to dismantle and dehumanize again (this theme of the self-regeneration of inhuman violence and horror seems particularly appropriate in the context of the US–Mexico border drug wars where drug groups and cartels often seem to have the uncanny ability to reform spontaneously[30]).

With Damiens' *supplice*, there was perhaps already a sense that early modern regimes of sovereign power and force could not fully apprehend the excess of violence on display. There was a beginning of an indication that agony and horror were possibly slipping out of the *dispositif* of ritualized and regulated (bio)political violence that had hoped to reveal the uncontested authority of a penal, legal, and governmental system (or what might have been modalities of biopolitical sovereignty in the making). Indeed, the horror of the criminal body's sacrifice continuing well beyond the death of the criminal opened the door for a "sovereign" (or at least, spontaneous and autonomous) force that no longer would have anything to do with any form of centralized and awe-inspiring power or with any attempt at disseminating governmental prerogatives to public agents (judges, penal officers, prison wardens, executioners, and so on) in charge of maximizing the life of the population or the human species. Foucault himself gives us some initial clues about this other sovereign force or violence. Foucault hints at the fact that the atrocious scenes of bodily dismemberments and obliteration of human dignity and individuality that took place in public repelled more and more audiences/members of the public who lost track of the intended awe-inspiring and deterring meaning of the *supplice*.[31] And as they lost track and started to question the meaning of the public visual *supplice*, they sometimes morphed into agents of a vengeful counter-justice or counter-power, one opposed to the state or to penal and governmental agents.[32] Some of the members of the public who violently reacted against the penal state revealed themselves eager to decapitate power and pulverize the sovereign's mastery (this, perhaps, culminated with the French Revolution, its literal decapitating of the sovereign's head, and the reign of *Terreur* that ensued). Thus, a counter-horror or a reverse inhumanity, much more popular this time, started to unfold and, as a result, forced the penal and judicial system to be far less visual and public (by removing the horror of the *supplice* from the public domain and by trying to bring the biopolitical power of justice and its agents into the more sedate domain of social disciplinization and normalization).

Once again, such an early modern counter-horror (even if, at times, it bore the name of *terreur*), in its opposition to the state's own violence, is not without recalling the contemporary drug cartels' limitless capacity for grotesque killing and dismantling of the human body. As the sovereign and its agents present themselves as the embodiment of a body politic, as the guarantors of the integrity of the social body, or as the maximizers or producers of a certain humanity, the singularity of human bodies and lives becomes a prime target of agonal sovereignty and of the agonal warriors' horror. Thus, it may be that agonal sovereignty, in either an early modern or a late modern context, is not that "post-political" after all. Although the pulverization of the flesh of the human does not require any specific political or ideological formula, and even though it may not be performed in the name of a new political order (for what conception of the political could accommodate itself of such an excessive violence and of such a surplus of inhumanity?), the (bio)political violence of the state and the state form is perhaps something that initially propels the agonal warriors or the enactors of the human sacrifice to unleash their arsenal of horrors. Thus, the horror-to-come or the inhumanity-to-be of agonal warriors, then and now, may also be an absurdly ironic reflection on modern political power, force, and violence, a modern power, force, or violence that visions and acts of horror push to the extreme, towards the fateful undoing of the human or human dignity that has been the justification for so many instances of political terrors over the ages. Here, we are once again reminded of Deleuze and Guattari's well known conceptualization of the war machine. Called for by political agents of the state (by the state form), the war machine turns into the accursed share of modern (bio)power and unleashes an agony of force, violence, and sovereignty that remains totally inaccessible to the state and its governmental agents (or what's left of them).[33] In an extreme twist of irony and absurdity, the war machine's grotesque horror meets the challenge of biopolitics and of modern (bio)power's own terrorizing violence.[34]

An agonal violence and horror in excess of biopolitics

As we stated in the "Introduction" to this volume, biopolitical frames of representability are epistemological frameworks that render intelligible practices of power and violence in terms of the preservation of life, a preservation of life that sometimes entails the possibility of taking away the lives or destroying the bodies of others (unworthy bodies or lives). We have argued throughout the preceding five chapters that this epistemological grid of intelligibility, as Foucault would call it, misrepresents or occludes instances of violence performed onto bodies and life, particularly excessive and horrific forms of agonal violence that, as we saw above with both the early modern and late modern *supplice*, appear to resist easy classifications or human comprehension. For example, even when they try to make sense of the manner in which Nazi totalitarianism operated by way of an unspeakable violence onto human bodies, biopolitical frames of representability, such as those developed by Agamben, Esposito, Dillon and Reid, or indeed Foucault, fall short of apprehending the condition of horror and agony characteristic of the camp (actual or virtual). Agamben sees the Nazi Lager as the culmination of a series of historical events tied to the production of bare life, often occurring under conditions of extreme sovereign decisionism too. For Foucault, the Nazi death machine represents the paroxysmal racialization of a biopower turned into a thanatopolitics of the human species. Both Agamben and Foucault remain caught within a dialectic of life or death. Even Esposito's claim that Nazism exemplifies the logic of immunitary responses to threats of bodily contagion with, along the way, an empowerment of the biomedical profession to decide who can live or die, or better yet, what is life and what is non-life, misses the specter of grotesque, gruesome, and ghastly horror that was in place in the Nazi laboratory of human sacrifice. Thus, it is not surprising to read that Cavarero calls many of these critical biopolitical perspectives "misleading" due to their keeping away from sight a violence that functions at a crucial ontological level, that of the unraveling of the human condition itself.[35]

For Cavarero (who turns to Arendt's analyses in *The Origins of Totalitarianism*), the camp represents a laboratory indeed, but not only in terms of the production of corpses. It is primarily a laboratory that devises techniques to annihilate the specific uniqueness of the human. As Cavarero argues: "horror has to do precisely with the killing of uniqueness … it consists in an attack on the ontological material that, transforming unique beings into masses of superfluous beings whose [quoting Arendt] 'murder is as impersonal as the squashing of a gnat,' also takes away from them their own death."[36] The destruction of human uniqueness or of the singular spontaneity of being (for Arendt, at least) often occurs by way of what Primo Levi has called a "useless violence," that is to say, a violence that appears to be beyond or, once again, in excess of the ultimate rationalization for organized (political) slaughter.[37] Levi's description of the process of dehumanization that defines and creates the *Häftlinge* (or detainees) of the Nazi system, from the initial boarding of the trains in the fog and the night (as is so graphically recounted by Alain Resnais in his film, *Nuit et Brouillard*[38]) to the inhuman labor conditions and myriad new rules (often unknown to the inmates) that dominated the Nazi camp, finally forces him to conclude: "One is truly led to think that, in the Third Reich, the best choice, the choice imposed from above, was the one that entailed the greatest affliction, the greatest waste, the greatest physical and moral suffering. The 'enemy' must not only die, he must die in torment."[39]

It is precisely this "human waste" dimension that remains "out of the frame" in biopolitical analyses of violence, war, and terror (past or present). While Agamben, Foucault, and Esposito (among others) develop a specific rationale for the emergence of genocidal possibilities in the twentieth century and beyond, they miss the crucial aspect of how a horror of excess and, similarly, an excess of horror account for what often gets defined (perhaps too hastily) as totalitarian forms of biopolitical governance. As Arendt also observed: "the real horror of the concentration and extermination camp lies in the fact that the inmates, even if they happen to keep alive, are more effectively cut off from the world of the living than if they had died, because terror enforces oblivion."[40] Thus, instead of insisting on the presence of a bifurcation between life and death, a biopolitical analytical threshold on the basis of which the promotion of the former and the inducement to the latter can be achieved, there is evidence of a certain praxis (whether it is "post-political" or not, as we said above, remains to be seen) that we have called agonal sovereignty. It is agonal sovereignty, not because it *must* be called this way, nor because the term offers a convenient symmetrical opposition to the idea of a biopolitical sovereignty advanced by some,[41] but because not calling it anything would likely contribute to forcing the horror back into biopolitical frames of representation that, once again, do not possess an adequate lens to shed the light on what is going on beyond the mere fact of life and death.

What is going on beyond the mere fact of life and death is not just one thing, or one modality of violence. It is in fact many instances of agony and horror that terms we have employed throughout this study, like excess, expenditure, pulverization, waste, or the scattered (*l'épars*), cannot claim to represent or explain fully. At best, such terms approximate or metaphorize a violence that, often, is simply unnameable (as Cavarero suggests). To reassert a point stated above (and theorized and illustrated in previous chapters), such an excess, expenditure, waste, surplus, or accursed and disseminated supplemental share operates on two complementary levels. As we saw above with the *supplice*, the excess (and other related figures) resides in the operationalization of violence itself. Horror is a persistent and apparently purposeless accumulation of violence onto the often already dead body, onto the corpse, whose death is no longer what needs to be assured. Thus, the agony of horror shares a resemblance with the fact of over-killing or over-destroying. Similar to the unmanned drone technologies of tracking and destruction employed by many militaries these days (including US troops in Iraq and Afghanistan), horror comes back several times over to steamroll the lifeless body, the dismantled human flesh, or the organs without bodies (an interesting reversal of the Deleuzian "bodies without organs" concept[42]).

But the overkill is not the only excess or surplus at work here. As we stated in Chapter 4, horrific violence also seeks to exceed the human, to surmount or overcome humanity. The figural unity that is incarnated by the human self and inscribed by his/her life (as an existential entity, as an *ontos*) is indeed surplussed by certain modalities of graphic and often visually appalling violence that, in lieu of the logic of the one, seek to propagate the logic of the multiple, of the "manifold" (as Deleuze would say), or of the multitudinally butchered and disseminated flesh. As we indicated above with the example of the *narcos*' violence in Mexico, horror's trend towards the overcoming of the human may also be a response—cynical, ironic, and abject, to be sure—to another, more conventional, more expected, and more modern (perhaps) violence, that of the state or state-form. Interestingly, this state-based terrorizing violence, over the ages, has rarely been seen as excessive or as falling on the side of horror (indeed, it is really not until the mid-twentieth century, in the aftermath of the Nazi genocides and with the advent of human rights discourses, that state or sovereign violence started to acquire a horrific dimension). Although we are not claiming to have found a justification for forms of violence that, in the end, resist human explanations and rationalizations (and that may just be their intention too), the possibility that the "anti-human" (to use Levi's term) dimension of horror may present an element of response to more conventional sovereign-centered or government-managed operations of violence, war, and terror cannot and probably should not be discounted. Again, returning to Georges Bataille's notions of the accursed share and of sacrificial economies briefly discussed at the end of Chapter 4, there is a daring, challenging, defiant, and reactive aspect to horror and horrific violence. This is perhaps also why horror's "political" status, its politics of agonal warring, is far from clear.

To the extent that they try to offer an answer or advance a modality of resistance to the violence that is both in excess of the kill and in excess of the human, proponents of biopolitical perspectives offer a rather predictable response or solution. Their answer seeks to remain within the (safe?) confines of biopolitical frames of representability and intelligibility. In a viciously circular way, this desire to preserve or reemploy biopolitical lenses also prompts them, consciously or not, to fold horror back within terror and to bring back the surplus of both life and death to an analytical terrain where the dialectical debate between a power to promote life and a power to put to death is seemingly endless. Crucially, by insisting on the timeless validity of biopower (whether they admit it or not), critical biopolitical theorists deprive themselves of the ability to perceive the fact that horror—gruesome, disgusting, and incomprehensible as it may well be—may also act as an extremely ironic and vengeful response to the modalities of power, force, and violence that they point to in their own analyses of the (late) modern condition of mobilization of and control over life and living bodies. One intriguing but, in the end, sadly futile biopolitical answer to horror (again, to the extent that critical biopolitical scholars venture at all into the domain of horror; most of them, as we established in the "Introduction," simply do not) consists in incanting the image or concept of an over-life or of over-living, a *survivance* (literally meaning "over" or "in surplus of life") that allegedly could serve as a buffer to the practices of over-killing and over-destroying of the human body. Thinkers like Bonnie Honig (drawing some of her insights from Jacques Derrida) and Judith Butler have advanced such a possibility.

In her latest book, Honig returns to some of Derrida's thoughts on survival and survivability (from his 1985 essay/interview "Deconstruction in America"[43]) to retrieve a notion of "more life" or "surplus life" (*survivance*, as Derrida puts it).[44] Over-living or *survivance* is a surplus of life that offers reasons to hope, even in the face of horrifying hostility, agonism, and violence. As Honig puts it, *survivance* as over-living is a "dividend—a surprise extra, the gift that exceeds rightful expectations, the surplus that exceeds causality."[45] But this additional gift of life, this more life, is not just about devising ways, while one is still alive, to maximize the conditions that may allow "mere life" (Honig's term) or even a qualified political life to last a bit longer, to be a bit stronger, or to save a bit more energy so as to try to fight off terror or horror a bit better (although terror or horror here, once again in a rather typical biopolitical fashion, become synonymous with the physical death of the body). Rather, *survivance*, as Honig interprets Derrida's concept, is also about "more than life." More than life, then, is "both *plus de vie* and *plus que vie*."[46] This "*plus que vie*" supplement suggests it is worthwhile for human beings to go on finding ways to over-live the overkill, because there is something about life as *survivance* that outlasts death or, as Honig would have it, can overcome extreme force and power found in "emergency settings."[47] Even when life is taken away from the body, even when a horrific violence insists on butchering or burning the lifeless corpse, *survivance* can still be present or available. *Survivance* is a surplus, but not only with regards to life. It is a surplus vis-à-vis horrific violence too. *Survivance* as "*plus que vie*" is that remainder that not even the most ghastly forms of destruction of the human flesh can target or capture. It is a surplus to that which the excess of violence has already managed to surplus (often the pulverized flesh). Put bluntly, according to Honig (and maybe Derrida as well), "*plus que vie*" can exceed the excess.

Thus, for Honig, the possibility of this surplus of all surpluses to be found on the side of an over-life (not to be confused, supposedly, with an after-life, although Honig is rather silent about the distinction between over-life and after-life) is a matter of continued trust in the power of and over life, in the life of the human, even if—or because—life has become a question of survival or survivability (and is no longer a matter of enjoying the political conditions offered by the "good life"). *Survivance*, she argues, leads us "toward overlife, toward the gifts of life, to the extra, the dividend, the unearned, and toward that which cannot be earned."[48] Recently, Didier Fassin has suggested that Derrida's "ethics of survival," allegedly affirmed by Derrida in his reflections towards the end of his life, follows a similar path since "[t]o survive is to be still fully alive and to live beyond death."[49] Fassin adds that *survivance* or life as survival is an "'unconditional affirmation' of life and the pleasure of living, and it is the hope of 'surviving' through the traces left for the living."[50] There is indeed, as Fassin intimates, and perhaps inevitably, a blind hope that seems to guide such a call to an over-life. Indeed, hope and fate appear to be, in Honig's theoretical construct and perhaps in Derrida's ethics of survival too, what ultimately *survivance* opens up to, all the more so since the dividend or remainder that is this life beyond life (and beyond death), this "*plus que vie*" that Derrida announces, "cannot be earned," as Honig once again claims. Since it cannot be earned, *survivance* either comes to happen or not, as a chance event perhaps, as a gift that some may come to enjoy or benefit from (although they may not know it since their bodily death may have already taken place and their body may have already been scattered about) and others may not.[51] This also begs the question of for whom or from what perspective is such an over-life of *survivance* important or meaningful? And does it have any bearing at all on the surplus or excess of horrific violence whose own eventness or emergence may have already targeted the body of the human? Indeed, Honig herself cannot be sure what the answers to such questions will be. She leaves us with questions about the *survivance* she incants, questions that may make us doubt the applicability of such a surplus of life and living in the face of another surplus, that of violence and horror that, to repeat, does not care to target life itself but rather seeks to disseminate the very notion of humanity. Indeed, Honig writes: "The question here then is what resources, concepts, and practices may promise survival as life and overlife, mere life and more life, to contemporary democracies?"[52] Perhaps Honig's main problem is that, like many other supporters of biopolitical arguments, she focuses on the wrong kind of unity or singularity: the unity of life and of its lasting (possibly eternal) capacity, as opposed to that unity or uniqueness that comes with the idea of humanity.[53] This misplaced conceptual and representational focus on what Rosi Braidotti calls "life as a self-constituting entity" may not allow Honig to see that what she hopes can be in excess (the mere life of the human) has indeed already been made excessive, residual, and superfluous.[54]

Judith Butler too has recently called for a return to the idea of survival or survivability. Reprising some of her earlier arguments about the "precariousness of life" under notions like "grievability" and "survivability" in her more recent essays,[55] Butler has suggested that such concepts can "call into question the ontology of individualism."[56] Interestingly, it appears that Butler's notions of precariousness, survivability, and grievability operate on the same ontological terrain as horror (as Cavarero and we, throughout this study, have tried to theorize it): that of the challenging of the unity, singularity, or uniqueness of human experience or existence. As such, Butler's thought may offer a challenge to horror by introducing another way, a less gruesomely destructive way, of dealing with the power and force of the one, with the omnipresence of the *ontos* (as we put it in the "Introduction" to this book). Butler still couches her analysis at the level of the body and the life of the body. But, in an age of terror and war, she urges us to reassess the body and to "reconsider the way in which we conceptualize the body in the field of politics."[57] Indeed, the singularity of the body has been misunderstood, Butler suggests. What makes the human body singular, individual, and of unique importance (the "body that matters," as she put it in her previous work[58]), is not physical characteristics but, rather, the condition of sociality with which every single body engages. As she puts it:

> [w]hat makes a body discrete is not an established morphology, as if we could identify certain bodily shapes or forms as paradigmatically human. In fact, I am not sure at all we can identify a human form, nor do I think we need to … We can think about demarcating the human body through identifying its boundary, or in what form it is bound, but that is to miss the crucial fact that the body is, in certain ways and even inevitably unbound—in its acting, its receptivity, in its speech, desire, and mobility. It is outside itself, in the world of others, in a space and time it does not control, and it not only exists in the vector of these relations, but as this very vector. In this sense, the body does not belong to itself.[59]

This dimension of "not belonging to itself" of the body opens it up (and all of us too) to a basic form of vulnerability or precariousness that, in Butler's perspective, is salutary. From such a fundamental opening, a social unity and solidity of the human body can be arrived at. Or, as Butler would have it, the lack of boundaries to the body in both space and time leads to a necessary sense of survival that results from the "constitutive sociability of the body," a constitutive sociability and survivability that can help to "minimize the conditions of precariousness in egalitarian ways."[60]

Butler's proliferation of the body, her opening up of human life to encounters with others and with the outside, seems to have the potential to operate analytically on a ground similar to that which we found with horror and agony. Indeed, both horror and Butler's precariousness do not require—in fact, they both resist—an ontological primacy given to the singular corporeality of the one human life (as the guarantee of human existence and survival).[61] Put differently, both Butler's precariousness or vulnerability and the condition of horrific violence challenge the premise that "the human form" should survive at all costs. Moreover, by placing her analysis of the precariousness of life in a conceptual domain where the body is always already understood as having been displaced, othered, unbound, and possibly disseminated across space and through time, Butler may hope to present a credible response to the spectacle of horrific violence that cannot care less for the body and life of the human. But, disappointingly, Butler does not follow through on such promising critical premises. Indeed, rather than engaging horror on its own terrain by placing it in contact with her own salvaging perspective on a proliferated body and life that owe nothing to "the human form," Butler instead falls back onto a politics of life and the living body, onto a form of biopolitics. While she argues that, as a result of encounters with an "outside" that is impossible to predict or plan for (and that may well be an outside or future of horrifying destruction too), certain "bodies will appear more precariously than others," she explains that it is indeed so because it all depends "on which versions of the body, or of morphology in general, support or underwrite the idea of the human life that is worth protecting, sheltering, living, mourning."[62] Thus, her explanations or justifications for why life or the body, at times, is more likely to be precarious (in other words, subject to the aleatory—and possibly violent—outside or future) has nothing to do with any notion of a possible chance encounter not just with death, but also (beyond death) with the horrifying event. Instead, the now selective precariousness of life and vulnerability of the body are the product of a "morphology" (the shape, look, context of the body), or of what mere life may always have in store for us (either surviving or succumbing). Again, this conclusion inevitably takes her back to a basic biopolitical frame of representability and intelligibility—an irony for one scholar so intent on resisting the "operations of the frame"[63]—whereby some "frameworks establish in advance what kind of life will be a life worth living, what life will be a life worth preserving, and what life will become worthy of being mourned."[64]

A few pages later in the same essay, Butler blatantly reveals her newfound hope (if not outright idealism) for what she ends up calling "stubborn life."[65] Studying the poems written by several Guantanamo inmates, poems by War on Terror detainees (some possibly survivors of the camp, others possibly dead and tortured) that, as she notes, have "survived the censorship of the US Department of Defense,"[66] she argues that those writings are signs of the presence of a surviving life. More crucially, the poems are life traces, a surplus of life, or an over-life even, as Honig might put it. They are marks of "a living being—a sign formed by a body, a sign that carries the life of the body."[67] These written leftovers or scattered bits that, we can surmise, both represent and parallel the leftovers and scattered bits of tortured and maybe pulverized bodies in the Guantanamo camp are, according to Butler, life beyond the body or life, a life that can thus survive outside of the body's own capacity for survivability. Clearly, even when confronted with an invisible mark or trace of the body (not even with bodily remains), Butler is unable to let go of life. Written traces such as those Guantanamo poems matter because bodies matter, because they are proxy bodily remains out of which the life of the Guantanamo inmates can be reconstructed or reinvented. As we suggested above in this volume, this idea of reconstructing the body, of producing a new life, is a strong biopolitical motif, one that Butler, desperate for the maintenance of "stubborn life" in the face of horror, finds herself replicating. And even if the body and life have been completely dismantled, even if the agonal warriors and horrifying butchers have already performed their ghastly deeds with regards to the human, there is still a life that can be rebuilt through survivability, precariousness, or vulnerability. There is still "grievable life … a life that can be regarded as a life, and be sustained by that regard."[68] There must still be signs or traces of something not just human, but also human life-like or humanly corporeal, in any fleshy bit, in any ash, or in any object and sign left behind or blown by the wind. Here, unlike Honig perhaps, Butler does not hesitate to merge the possibility for over-living with the hope for an after-life or for eternal living, since something, even if only words, will have to "endure beyond us, outlast our breath, break the shackles of solitude, transcend our transitory body."[69] With the help of the Guantanamo poems, Butler thus wishes to restore the precariousness or grievable potential of the "unbound body" not just to a biopolitics of new life creation or regeneration, but also, in case such a newly biopolitically produced life were not enough, to the affirmed ever present possibility of transcendent life.[70] Any thought, any ideal that is about life, living, and the live body must be urgently recalled, it appears, no matter how desperate and no matter how pointless, in order not so much to fight off horror and its violence, but rather to continue to hope that such a horror will never be able to take us beyond the fact of life or death.

Coping with horror and the dismantling of the human

As we showed in the "Introduction" to the volume with our discussion of Peter van Agtmael's war photography, and as seems to be confirmed by Butler's recent analytics of precariousness or grievability of life in the face of dehumanizing violence, biopolitical frames of representability are driven by a desire to legitimize the preservation of life. Such an analytical drive forces proponents of critical biopolitical perspectives to maintain life or the life of the body (bare life, mere life) as the central focus of analyses intent on making sense of practices of political violence, including war and terror. But, as we explained above and as we revealed throughout the chapters in this volume, biopolitical frames of representation have little to say or show about horror and about the agonal free-range that its practitioners enjoy. As a consequence, they may not have much to offer in terms of possibilities of resistance to horror, to agonal sovereignty, and to the agonal warriors' challenging and defiant anti-human violence. Must we conclude, then, that we are *beyond biopolitics* not only theoretically or conceptually, but also because it is indeed pointless, it is beyond any human capacity, beyond any thought, to challenge horror's own challenge? Is there no way to counteract the propensity for proliferation and scattering of agonal warriors and butchers beyond biopolitics? Are we doomed to the excess of horrific violence, both in terms of its overkilling and in terms of its surmounting of the human?

Likewise, and more to the point with regards to our own writing and theorizing in this book, can we present, linguistically or visually, the horror perpetrated by today's (but also yesterday's and probably tomorrow's) agonal sovereigns without falling back into a modality of normalization, categorization, rationalization, or justification that inevitably recreates incomplete frames of vision and representability? In other words, can we foster or encourage a capacity to continue to think in the face of horror, an ability to (as Arendt puts it) "*stop*-and-think,"[71] on the part of us, the spectators or witnesses that, so far, may not yet have seen our humanity pulverized by war machines, hunting butchers, or recreational executioners? Can horror allow us to remain champions of critical thinking whereby thinking, as Michael Shapiro so lucidly puts it, "involves resistance to the dominant modes of representing the world, whether those representational practices function as mere unreflective habit or as intentionally organized, systematic observation."[72]

To a certain extent, philosopher Cora Diamond is puzzled by similar questions. In her essay "The Difficulty of Reality and the Difficulty of Philosophy," Diamond's focus is apparently on the way analytical philosophy remains incapable of articulating any notion of responsibility towards nonhumans.[73] But her more general argument revolves around the manner in which philosophy transforms conditions of human vulnerability into moral and ethical issues or problems. In the course of developing her argument, Diamond turns to the notion of horror. As she argues, the knowledge of horror—in her case, the horror felt by Elizabeth Costello, the main protagonist of J. M. Coetzee's novel *The Lives of Animals*, with respect to the chain of food production that Costello equates to the Nazi Holocaust—"pushes us beyond what we can think. To attempt to think it is to feel one's thinking come unhinged."[74] Diamond adds that "[o]ur concepts, our ordinary life with our concepts, pass by this difficulty as if it were not there; the difficulty, if we try to see it, shoulders us out of life, is deadly chilling."[75] Philosophy for

Diamond (who draws her insights from Stanley Cavell's skepticism) does not "appreciate a difficult reality," namely, the capacity of understanding the wounded and isolated Costello's inability to deal with and articulate the horror she is faced with. Indeed, Diamond goes on to state that philosophy deflects "our own capacity to inhabit in imagination the body of a woman confronting, trying to confront, what we do to animals."[76] Put differently, philosophy deflects our capacity to face horror by seeking to translate the issue of confronting that which horrifies us into a moral question. Diamond adds that "philosophy does not know how to inhabit a body (does not know how to treat a wounded body as anything other than a fact)."[77] Being cognizant of the precariousness, vulnerability, or fleshiness of others (human or animal bodies, for Diamond) can provoke both an incredible panic (and an incapacity to move or act, as Cavarero already told us about horror) and an isolation/loneliness that often render the turn to abstract moral reasoning a necessary lure. Again, it may be that, at the end of their analysis, both Honig and Butler fall victim of such a lure too.

Diamond's reflections raise for us another important interrogation. It is an interrogation that, to some, may be an endpoint for thinking but that, instead, can be seen as a necessary interruption of the often obvious turn towards an ethics of responsibility or care for the grievable or vulnerable body/life (an ethical turn or transformation that, once again, can sometimes force a closure for any critical thinking confronted with horror). Quoting Cavell, Diamond writes that philosophy must be "turned equally toward splendor and toward horror, mixing beauty and ugliness; turned toward before and after; toward flesh and blood."[78] Critical thought (what philosophy could enable) can expose the language-games and the image-games of horror and agony in all of their gruesome provocation. But critical thinking can also seek to provide responses to agonal sovereignty's horrific violence, responses that need to be ever more thought-provoking and conceptually provocative. Here, Diamond introduces the possibility that thinking can function as a challenge to horror. Still, thinking as a challenge to horror comes at a cost or with a burden: that of embracing not so much horror and its dehumanization, but the defiant provocation that makes the scene or event of horror so shocking, paralyzing, and unexplainable. Put differently, it is at the level of provocation—provocation of thought, affect, understanding, and visual perspective—that critical thinking needs to be placed if it is to confront horror's own representational defiance.

One possible way for thought to continue to be provocative and provoking is to push forward another ethic, an excess of ethic, in a way. Mirroring the excess of the flesh of the multitude achieved by horror's violence, this ethic is not one that, as suggested above, longs for the care of the vulnerable body or seeks to reinvent moral ways of remembering or reliving grievable life. Rather, it can be formulated as *an ethic of inventorying the scattered.*[79] By marking various episodes of horrific violence and by remarking its excess of both the human and the violent kill, such an ethic of inventorying the scattered offers several challenging possibilities. First, by being dispersed, by tracking some body parts, bits of flesh, ashes and other human or non-human remains, it does not claim to be thorough, exhaustive, logical, linear, or even constructive. This ethic of inventorying the scattered advances a mode of visual, linguistic, and conceptual apprehension that is as scattered as the matter it inventories. As such, it resists the drive to reconstruct the body or life, even a body beyond the body or a life beyond life. Restoring a unity or a uniqueness, whether human or not, is in no way the objective of such an ethic of inventorying the scattered. Second, through its inevitably disjointed unfolding, this ethic resists the need to categorize, define, or totalize horror and its effects or affects. It cannot wish to return the violence that reduces the human to ashes to be dispersed by the wind to previously established analytical frames, be they biopolitical or otherwise. At best, as we indicated above, it offers a few terms, a few concepts, a few visual signposts along the way, to draw our attention towards that which has been pulverized, but not to label it or to visually reframe it.[80] Inventoried in the course of the exploration, the names, concepts, and visual signposts that are being tentatively introduced have no hope of being memorialized or of turning into foundations for any new grand theorizing. Again, these names, images, or conceptual possibilities are principally mirror effects or mimetic forms of the scattered that, as such, may be able to open up additional challenges for critical thinking.[81] Third, the ethic of inventorying the scattered resists the all too common impulse to closure in the face of horror, something Diamond suggested above. Confronted with horrifying sights, facts, descriptions, or sentiments that are beyond what is humanly graspable, thought—particularly ethical thought—has a tendency to recoil, to shrink, or to rally around the mark of the human, around human life, or around the idea of human integrity or identity. By contrast, the ethic of inventorying the scattered offers openness as a counterpoint to horror's gaping hole or wound. It offers multiplicity of thought, of reasons, of explanations for what may or may not have happened to the body targeted way beyond the fact of life and death. It enables an opening, a space to rethink relations between what is alive and what is not and, perhaps just as crucially, between what may be alive or always subject to death and what does not account for either life or death. This open possibility to rethink relations not so much between various forms of life (or non-life) but also between modalities of apprehension of the materiality of the flesh, fleeting and fragile as such a materiality may be, is possibly something that Butler was trying to gesture towards in her own attempt at reconsidering the notion or even appearance of the body. Such an open space made available to explore new relationalities between objects and subjects under conditions not so much of *survivance*, but of emergent (in)humanity (or of an inhumanity-to-come, as we argued in Chapter 4) may be seen by some as a new political space.[82] While it may be perceived as a way of rethinking the condition of possibility for the political in times of horror and agonal warring, the recreation of a politics or the regeneration of the political is not an objective of the ethic of inventorying the scattered.[83] Rather, what we are suggesting here is that the opening up of the space itself—and the ethical call for the continued opening of this space through the challenge of critical thinking—may be what is political, what politics becomes, in such an ethic of openness in the face of horror.[84]

A difficulty possibly encountered by this ethic of inventorying the scattered is the fact that language has a limited capacity to capture horror. Here, it may be useful to remember Cavarero's point that horrific violence often manifests itself as a visual scene, as a sight that seizes the gaze and defies pre-existing rationalizations typically dependent upon linguistic constructs and narrative frames (or on prior visual representations already legitimized by and enshrined in linguistic contexts).[85] We may also recall the conceptual difficulty encountered by Foucault in *Discipline and Punish* when he narrated the part of Damiens' *supplice* that targeted the body beyond the moment of the condemned's death. Horror seems to be able to expose the communicative shortcomings of language. Or, to put it another way, horror reveals the aporia of systems of meaning and representation based on logics of communication. Thus, any ethic of inventorying the scattered needs to contend with the possibly thwarted capacity to bear witness, particularly to the extent that witnessing requires narrative coherence or unity. But there is a crucial distinction to be made between bearing witness and inventorying the scattered. As some have argued, bearing witness is often premised upon narrative coherence. Bearing witness seeks to reconstruct a vocabulary and retrieve an organization of concepts that can help us make sense of the so-called unspeakable savagery.[86] Storytelling is thus essential to bearing witness. Inventorying the scattered, by contrast, has no need for such a narrative unity or for representational consistency, as we explained above. Inventorying the scattered deals with a messy field, the same messy arsenal of horrors that bearing witness allegedly faces too. But inventorying the scattered is not obsessed by the requirement to make sense or by the need to verbalize in order to better learn lessons about the past or rationalize about the future. Thus, exposing or displaying loosely connected scenes of horror is not something that troubles the ethic of inventorying the scattered because such an ethic, to repeat, does not demand that it be translated into a larger, and supposedly morally or humanly coherent, system of thought or representation. In fact, it may be that the ethic of inventorying the scattered is better suited to cope with horror's traumatic fragments because its raking up of loose bits of visual information (if not meaning) here and there from the fields of horror and agony prepares it well to navigate throughout the visual domain where, as Cavarero once again intimates, horror propagates its most paralyzing and destructive effects.

Last but not least, the ethic of inventorying the scattered maintains and proliferates a capacity for provocation that is crucial to thinking about horror (and to critical thinking, in general). It allows us to see the perverse possibility about horror that we started to sketch above: namely, the possibility that horror is the image or scene of *our own* undoing and sacrifice. By refusing to frame horror within a particular configuration of violence or brutality to which we must give a name or for which we must provide rational explanations, the ethic of inventorying the scattered displays the sacrificial relinquishing of our humanity, of our human and humanistic values (including those of life and bodily integrity), perpetrated by ourselves, by our own excess of humanity, or by the human species' need or desire to burn out its own moral, ideological, and political surpluses. In one of his last theoretical provocations, Jean Baudrillard offers a perspective that comes close to the challenge of the ethic of inventorying the scattered we are seeking to present here. Baudrillard writes:

> It is everything by which a human being retains some value in his own eyes that we (the West) are deliberately sacrificing. *Our* potlatch is one of baseness, shamelessness, obscenity, debasement, and abjection. This is the whole movement of our culture—it is here that we raise the stakes … Our potlatch is the potlatch of indifference—an in-differentiation of values, but also an indifference to ourselves … And it is this indifference and abjection that we throw out to the others as a challenge: the challenge to debase themselves in their turn, to deny their own values, to lay themselves bare, to make their confessions, to own up—in short, to respond with a nihilism equal to our own.[87]

While Baudrillard's main target is specifically Western modernity and Western modern values, there is a sense in his analysis that humanity (as it has been invented and preserved throughout modernity) is not only capable but in fact eager to provoke its own sacrificial horror. As Georges Bataille had already theorized, our humanity is always already productive of the sacrificial excess,[88] and it is always already open to the condition of horror. Thus, it is far too simplistic and convenient, the ethic of inventorying the scattered periodically reminds us, to assume that there must be a duality between horror and us, or between horror and life (even if life is mere life or bare life). Pursuing some of Bataille's and Baudrillard's own defiant themes, the ethic of inventorying the scattered pushes us to question why it is that we must assume that humanity is an "us" and horror is a "them" (or that horror must be presented as other to us). While such a basic duality may be required to sustain a sense of the importance of life (and of the figural unity of the human incarnated by the individual body), it does not stand the test of the repertory of horrific violence performed by the ethic we are offering here, an ethic that adumbrates any and all horrors, even those that are the results of our own self-immolation (and this includes some of the horrors or terrors that, as we mentioned above, typically have not been connected with the condition of horror, mostly because they were said to be legitimized or authorized by the state-form or by political notions of sovereignty or governance). Thus, the ethic of inventorying the scattered may well be performing an important witnessing function too, even if it is one that is quite different from the call to bear witness. Indeed, the ethic of inventorying the scattered is a casual witness to the "theater of our own horrific decomposition."[89] Far from seeking to resist or confront horror and agony with analytical tools borrowed or recycled from other systems of thought or morality, the ethic of inventorying the scattered makes itself available for more conceptual challenges and provocations vis-à-vis horrific violence, a violence that undoes our humanity and yet is so common to it. It embraces an openness of thinking, perhaps a virtuality of thinking (as we suggested in Chapter 3), that is our human, non-human, or inhuman—as Baudrillard intimates, we have grown indifferent to the distinction between those concepts—best chance to cope with the unexpected excess that, for good or bad, is to come. It prepares us as best as can be for a condition, a violent surplus about who we are or, better yet, who we claim to be that is as aleatory, uncertain, and scattered as it is often horrific. After all, as Baudrillard prophesizes: "[e]ven the most extreme, most sublime thing we can conceive of will be taken over and surpassed by some other form—perhaps even by its opposite or caricature. This is how it is … Matters are never definitively settled."[90]

Notes

Introduction

1 Susan Sontag, *Regarding the Pain of Others* (New York: Farrar, Strauss and Giroux, 2003), p. 14.
2 Roberto Esposito, *Bios: Biopolitics and Philosophy*, trans. Timothy Campbell (Minneapolis: University of Minnesota Press, 2008), p. 9.
3 P. W. Singer, *Wired for War: The Robotics Revolution and Conflict in the 21st Century* (New York: Penguin, 2009), p. 436.
4 Judith Butler, *Frames of War: When is Life Grievable?* (New York: Verso, 2009), pp. 63–100.
5 Ibid., p. 67. See also Sontag, *Regarding the Pain of Others*, p. 89.
6 Butler, *Frames of War*, p. 67.
7 Ibid., p. 67.
8 Ibid., p. 68.
9 Ibid., p. 68.
10 Ibid., p. 68. See also Sontag, *Regarding the Pain of Others*, p. 18.
11 Butler, *Frames of War*, p. 69. See also Sontag, *Regarding the Pain of Others*, pp. 68–69.
12 Butler, *Frames of War*, p. 71.
13 Ibid., p. 71.
14 Peter van Agtmael, *2nd Tour, Hope I don't Die* (Portland, OR: Photolucida, 2009).
15 Ibid., p. 3.
16 Sontag, *Regarding the Pain of Others*, pp. 26–27.
17 Van Agtmael, *2nd Tour*, p. 3.
18 Ibid., p. 3.
19 Ibid., p. 3.
20 Michel Foucault, *The History of Sexuality*, Vol. 1 (New York: Vintage Books, 1978), pp. 138–39.
21 Van Agtmael, *2nd Tour*, p. 3.
22 Giorgio Agamben, *Homo Sacer: Sovereign Power and Bare Life* (Stanford, CA: Stanford University Press, 1998), p. 2.
23 Ibid., p. 3.
24 Ibid., p. 8.
25 See also Julian Reid, *The Biopolitics of the War on Terror: Life Struggles, Liberal Modernity, and the Defense of Logistical Societies* (Manchester: Manchester University Press, 2006). Reid argues that lives and bodies are "deployed biopolitically" in today's war on terror (see Reid, *The Biopolitics of the War on Terror*, p. 33). It appears that van Agtmael's visual frame seeks to reflect/embody such a claim.

26 Butler, *Frames of War*, p. 17.
27 Notable exceptions are two photos. One depicts the body of a presumed Iraqi insurgent, with his head partially blown as a result of a firefight with US troops. The other, on the adjacent page, shows another Iraqi, detained after a raid by a US patrol on a Mosul house thought to hide a torture room, with his face covered with a scarf. It turns out that this Iraqi individual was an informant. Whereas bodies and faces of US soldiers (or allied Afghan and Iraqi troops) and of a few Iraqi and Afghan civilians (often appearing to be innocent victims in the war) are on display, those of the "enemies" are by and large occluded.
28 Michael Dillon and Julian Reid, *The Liberal Way of War: Killing to Make Life Live* (London: Routledge, 2009), p. 20.
29 Van Agtmael, *2nd Tour*, p. 3.
30 Michel Foucault, *"Society Must Be Defended": Lectures at the College de France, 1975–76* (New York: Picador, 2003), p. 241.
31 Nikolas Rose, *The Politics of Life Itself: Biomedicine, Power, and Subjectivity in the Twenty-First Century* (Princeton, NJ: Princeton University Press, 2007), p. 54.
32 Foucault, *The History of Sexuality*, Vol. 1, p. 143.
33 Rose, *The Politics of Life Itself*, p. 53.
34 See Michel Foucault, *Discipline and Punish: The Birth of the Prison* (New York: Vintage Books, 1979).
35 Rose, *The Politics of Life Itself*, p. 53.
36 Ibid., p. 54.
37 Foucault, *The History of Sexuality*, Vol. 1, p. 137.
38 Michael Dillon, "Security, Race and War," in Michael Dillon and Andrew Neal (eds), *Foucault on Politics, Security and War* (New York: Palgrave, 2008), p. 175; author's emphasis.
39 Rose, *The Politics of Life Itself*, p. 52.
40 For a seminal exposition of the inside/outside problematique, see R.B.J. Walker, *Inside/Outside: International Relations as Political Theory* (Cambridge: Cambridge University Press, 1993).
41 Dillon, "Security, Race and War," p. 195.
42 Ibid., p. 195.
43 Ibid., p. 168.
44 Achille Mbembe, "Necropolitics," *Public Culture*, Vol. 15, No. 1 (2003), p. 39.
45 Foucault, *"Society Must Be Defended"*, p. 254.
46 Dillon, "Security, Race and War," p. 170.
47 Catherine Mills, "Biopolitics, Liberal Eugenics, and Nihilism," in Matthew Calarco and Steven DeCaroli (eds), *Giorgio Agamben: Sovereignty and Life* (Stanford, CA: Stanford University Press, 2007), pp. 186–87.
48 Mbembe, "Necropolitics," p. 40.
49 Dillon, "Security, Race and War," p. 169.
50 See Foucault, *The History of Sexuality*, Vol. 1.
51 Ibid., p. 167.
52 Agamben, *Homo Sacer*, p. 122.
53 This is precisely why, according to Agamben, the threshold between what counts as meaningful life and what does not, the so-called inside/outside divide, constantly shifts and ceaselessly needs to be redefined. More than the divide itself, it is the constant shift between life and death, inclusion and exclusion, friend and foe that is the main concern of contemporary biopolitical thinking and action, and it is *homo sacer* that bears the marks of such a powerful indecision or indistinction. See ibid., p. 131.
54 Esposito, *Bios*, p. 9.
55 Ibid., p. 39.
56 Ibid., p. 46.

57 Ibid., p. 32.
58 Dillon and Reid, *The Liberal Way of War*, p. 23.
59 Esposito, *Bios*, p. 116.
60 Dillon and Reid, *The Liberal Way of War*, p. 125.
61 Ibid., p. 125.
62 Mbembe, "Necropolitics," p. 12. Some examples of this recent proliferation of studies advancing a biopolitical perspective are (in no particular order): Reid, *The Biopolitics of the War on Terror*; Elizabeth Dauphinee and Cristina Masters (eds), *The Logics of Biopower and the War on Terror* (London: Palgrave, 2006); Jenny Edkins, Veronique Pin-Fat, and Michael J. Shapiro (eds), *Sovereign Lives: Power in Global Politics* (New York: Routledge, 2004); Mark Duffield, *Development, Security, and Unending War: Governing the World Peoples* (Cambridge: Polity, 2007); Nicole Shukin, *Animal Capital: Rendering Life in Biopolitical Times* (Minneapolis: University of Minnesota Press, 2009); Stephen Morton and Stephen Bygrave (eds), *Foucault in an Age of Terror: Essays on Biopolitics and the Defence of Society* (London: Palgrave, 2008); Jeremy Youde, *Biopolitical Surveillance and Public Health in International Politics* (London: Palgrave, 2010); David O' Kane and Tricia Redeker Hepner (eds), *Biopolitics, Militarism, and Development: Eritrea in the Twenty-First Century* (Oxford: Berghan Books, 2009); Derek Gregory, "The Biopolitics of Baghdad: Counterinsurgency and the Counter-City," *Human Geography*, Vol. 1 (2008), pp. 6–27; Kolson Schlosser, "Bio-Political Geographies," *Geography Compass*, Vol. 2, No. 5 (2008), pp. 1621–34; Michael Dillon, "Governing through Contingency: The Security of Biopolitical Governance," *Political Geography*, Vol. 26, No. 1 (2007), pp. 41–47; Michael Dillon and Luis Lobo-Guerrero, "Biopolitics of Security in the 21st Century: An Introduction," *Review of International Studies*, Vol. 34, No. 2 (2008), pp. 265–92; Lauren Martin, "Bombs, Bodies, and Biopolitics: Securitizing the Subject at the Airport Security," *Social and Cultural Geography*, Vol. 11, No. 1 (2010), pp. 17–34; David Campbell, "The Biopolitics of Security: Oil, Empire, and the Sports Utility Vehicle," *The American Quarterly*, Vol. 57, No. 3 (2005), pp. 943–72; Claudio Minca, "Giorgio Agamben and the New Biopolitical Nomos," *Geografiska Annaler*, Vol. 88, No. 4 (2006), pp. 387–403; Paolo Palladino, "Revisiting Franco's Death: Life and Death and Biopolitical Governmentality," in Michael Dillon and Andrew Neal (eds), *Foucault on Politics, Security and War* (London: Palgrave, 2008), pp. 115–31; Paul Patton, "Agamben and Foucault on Biopower and Biopolitics," in Matthew Calarco and Steven DeCaroli (eds), *Giorgio Agamben: Sovereignty and Life* (Stanford: Stanford University Press, 2007), pp. 203–18; Julian Reid, "The Biopolitics of American Security Policy in the Twenty-First Century," in François Debrix and Mark J. Lacy (eds), *The Geopolitics of American Insecurity: Terror, Power and Foreign Policy* (London: Routledge, 2009), pp. 126–42.
63 Foucault, *"Society Must Be Defended"*, p. 242.
64 See Rose, *The Politics of Life Itself*, p. 54.
65 Adriana Cavarero, *Horrorism: Naming Contemporary Violence* (New York: Columbia University Press, 2007), p. 2.
66 Michael Hardt and Antonio Negri, *Multitude: War and Democracy in the Age of Empire* (New York: Penguin, 2004).
67 Ibid., p. 12.
68 Ibid., p. 13.
69 Ibid., p. 14.
70 Reid, *The Biopolitics of the War on Terror*, p. 17.
71 Ibid., p. 124.
72 Ibid., p. 124.
73 Ibid., p. 18.

74 Cavarero, *Horrorism*, p. 5.

75 Ibid., p. 5.

76 From this perspective, the war on terror did and still does display moments of terror, for example, right after the collapse of the Twin Towers in Manhattan on 9/11 or as a result of the advance of US forces into Baghdad in March 2003. But, Cavarero would suggest, not everything in the war on terror was/is actually about sudden movement or escape as a result of fear.

77 Cavarero, *Horrorism*, p. 7.

78 Ibid., p. 8. Here, Cavarero's theorization of horror comes very close to Julia Kristeva's concept of abjection. But Kristeva's abjection also insists on the element of attraction or fascination that is as much a part of disgust as is the desire to turn away form the horrifying scene, something that Cavarero does not spend too much time discussing. See Julia Kristeva, *Powers of Horror: An Essay on Abjection* (New York: Columbia University Press, 1982).

79 Cavarero, *Horrorism*, p. 8; our emphasis.

80 Ibid., p. 43.

81 Ibid., p. 9.

82 See Primo Levi, *Si c'est un Homme* (Paris: Pocket Juliard, 2008), p. 307; our translation.

83 Ibid., p. 310; our translation.

84 Ibid., p. 310; our translation.

85 We are indebted here to William Connolly's qualification of ontology as the study of the fundamental logic of the one existing thing or the one "design of being." See William Connolly, *The Ethos of Pluralization* (Minneapolis: University of Minnesota Press, 1995), p. 1.

86 Cavarero, *Horrorism*, p. 9.

87 Rosi Braidotti notes that, by contrast, biopolitics or biopower "has emerged as an organizing principle for the proliferating discourses and practices that make technologically mediated 'life' into a self-constituted entity." See Rosi Braidotti, "The Politics of 'Life Itself' and New Ways of Dying," in Diana Coole and Samantha Frost (eds), *New Materialisms: Ontology, Agency, and Politics* (Durham, NC: Duke University Press, 2010), p. 201.

88 See Jane Bennett, *Vibrant Matter: A Political Ecology of Things* (Durham, NC: Duke University Press, 2010), p. 23.

89 Writings by Arendt that Cavarero turns to in her study. See ibid., pp. 40–46. See, in particular, Hannah Arendt, *The Human Condition* (Chicago: University of Chicago Press, 1958), and Hannah Arendt, *The Origins of Totalitarianism* (New York: Harvest Brace Jovanovich, 1951).

90 Or, to use Cavarero's turn of phrase, because "the boundary between life and death has collapsed." See Cavarero, *Horrorism*, p. 42.

91 It would be interesting to consider what van Agtmael's photos from the wars in Iraq and Afghanistan would have looked like if he had not privileged a field of vision dominated by a biopolitical understanding or mode of representability, or if he had allowed horror to enter the pictures.

92 Here, we understand ontopolitics in a manner congruent with the way Bill Connolly and David Campbell have introduced the concept. Loosely articulated, ontopolitics refers to a representation/interpretation of politics (or political life) that establishes *fundamental* claims about actuality or presence without which the centrality, singularity, and uniqueness of being could not be established. Likewise, Connolly specifies, to deny such fundamental claims is also to partake of ontopolitics. As Connolly puts it: "To say either that something is fundamental or that nothing is fundamental, then, is to engage in ontopolitical interpretation." See Connolly, *The Ethos of Pluralization*, p. 1. See also David Campbell, *National Deconstruction: Violence, Identity, and Justice in Bosnia* (Minneapolis: University

of Minnesota Press, 1998), p. 22. Put differently, ontopolitics seeks to establish or situate the logic of the one "really existing thing" (the *ontos*) into a fundamental placing, one that will "fix possibilities, distribute explanatory elements, generate parameters within which an ethic is elaborated, and center … assessments of identity, legitimacy, and responsibility." See Connolly, *The Ethos of Pluralization*, p. 2. A traumatic or disastrous ontopolitics, then, would be one mode of representation or interpretation of political possibilities that does not so much establish or deny fundamentals but rather, problematically for the *ontos*, destroys or indefinitely postpones the very idea or image of a fundamental unity or identity (or of the singular, once again).

93 For a radically different perspective that seeks to champion the idea of the flesh of "the multitude" of bodies and lives, or what they hope will amount to a new "social flesh … [that] is pure potential … [and is] aimed constantly at the fullness of life," see Hardt and Negri, *Multitude*, p. 192. The flesh we are witnessing in moments of horror and through agonal sovereignty, by contrast, precisely seeks to obliterate this so-called fullness of life and humanity.

94 This possibility of ontopolitical action in the form of agonal sovereignty is perhaps not so surprising since, after all, as Connolly explains, it is often the case that the "ontopolitical stance … breaks with … ontological traditions." See Connolly, *The Ethos of Pluralization*, p. 5.

95 See Gilles Deleuze and Felix Guattari, *A Thousand Plateaus: Capitalism and Schizophrenia* (Minneapolis: University of Minnesota Press, 1994), pp. 351–423 and pp. 474–500.

96 For a previous study that initiated this line of thinking and introduced the concept of agonal sovereignty, see François Debrix, *Tabloid Terror: War, Culture, and Geopolitics* (New York: Routledge, 2008), pp. 93–99 and pp. 111–17.

97 See, for example, Hardt and Negri, *Multitude*. See also Michael J. Shapiro, *Methods and Nations: Cultural Governance and the Indigenous Subject* (New York: Routledge, 2003), pp. 181–82, and Debrix, *Tabloid Terror*, pp. 104–10.

98 See, in particular, Dana Villa, *Politics, Philosophy, Terror: Essays on the Thought of Hannah Arendt* (Princeton, NJ: Princeton University Press, 1999); Lisa Disch, *Hannah Arendt and the Limits of Philosophy* (Ithaca: Cornell University Press, 1994); and Bonnie Honig, *Political Theory and the Displacement of Politics* (Ithaca: Cornell University Press, 1993), pp. 76–96.

99 This is Žižek's formulation with regards to the consequences of Carl Schmitt's sovereign decision on the exception, a sovereign decision that, as we show in Chapter 1, is drastically opposed by Arendt's political thought, but whose effects (such as the unleashing of unrestrained war violence) may be oddly similar to some of the consequences of Arendt's agonal action. See Slavoj Zizek, "Carl Schmitt and the Age of Post-Politics," in Chantal Mouffe (ed.), *The Challenge of Carl Schmitt* (London: Verso, 1999), p. 18.

100 For some recent critical analyses of these contemporary issues and topics, see, for example, Kam Shapiro, *Sovereign Nations, Carnal States* (Ithaca: Cornell University Press, 2003); Michael J. Shapiro, *Methods and Nations: Cultural Governance and the Colonial Subject* (London: Routledge, 2003); Derek Gregory, *The Colonial Present: Afghanistan, Palestine, Iraq* (Oxford: Blackwell, 2004); Jenny Edkins, Michael J. Shapiro, and Veronique Pin-Fat (eds), *Sovereign Lives: Power in Global Politics* (London: Routledge, 2004); eds. Bulent Diken and Carsten Laustsen, *The Culture of Exception: Sociology Facing the Camp* (London: Routledge, 2005); Stuart Croft, *Culture, Crisis and America's War on Terror* (Cambridge: Cambridge University Press, 2006); Wendy Brown, *Regulating Aversion: Tolerance in the Age of Identity and Empire* (Princeton, NJ: Princeton University Press, 2006); Rey Chow, *The Age of the World Target: Self-Referentiality in War, Theory, and Comparative Work* (Durham, NC: Duke

University Press, 2006); Derek Gregory and Alan Pred (eds), *Violent Geographies: Fear, Terror, and Political Violence* (London: Routledge, 2006); Jean-Claude Paye, *Global War on Liberty* (New York: Telos Press, 2007); Ashley Dawson and Malini Johar Schueller (eds), *Exceptional State: Contemporary US Culture and the New Imperialism* (Durham, NC: Duke University Press, 2007); Anthony Burke, *Beyond Security, Ethics and Violence: War against the Other* (London: Routledge, 2007); Christopher Coker, *The Warrior Ethos: Military Culture and the War on Terror* (London: Routledge, 2007); Mark Neocleous, *Critique of Security* (Montreal: McGill-Queen's University Press, 2008); Sheldon Wolin, *Democracy Incorporated: Managed Democracy and the Specter of Inverted Totalitarianism* (Princeton, NJ: Princeton University Press, 2008); Danilo Zolo, *Victor's Justice: From Nuremberg to Baghdad* (London: Verso, 2009); Stuart Elden, *Terror and Territory: The Spatial Extent of Sovereignty* (Minneapolis: University of Minnesota Press, 2009); Bonnie Honig, *Emergency Politics: Paradox, Law, Democracy* (Princeton, NJ: Princeton University Press, 2009); eds. Didier Bigo and Anastassia Tsoukala, *Terror, Insecurity and Liberty: Illiberal Practices of Liberal Regimes after 9/11* (London: Routledge, 2009); François Debrix and Mark J. Lacy (eds), *The Geopolitics of American Insecurity: Terror, Power, and Foreign Policy* (London: Routledge, 2009); Nick Vaughan-Williams, *Border Politics: The Limits of Sovereign Power* (Edinburgh: Edinburgh University Press, 2009); and Dillon and Reid, *The Liberal Way of War.*

1 Agonal sovereignty: rethinking war and politics in an age of terror

1 Carl Schmitt, *The Concept of the Political* (Chicago: University of Chicago Press, 1996), p. 34.

2 See, for example, David Harvey, *The New Imperialism* (Oxford: Oxford University Press, 2005); Neil Smith, *The Endgame of Globalization* (New York: Routledge, 2004); or Derek Gregory, *The Colonial Present: Afghanistan, Palestine, Iraq* (Oxford: Blackwell, 2004).

3 See Michel Foucault, *The History of Sexuality*, Vol. 1 (New York: Vintage Books, 1988).

4 Among recent international relations or geopolitical studies that have focused on biopolitics and abuses of sovereign authority after 9/11, see Cristina Masters and Elizabeth Dauphinee (eds) *The Logics of Biopower and the War on Terror: Living, Dying, Surviving* (New York: Palgrave, 2006); Julian Reid, *The Biopolitics of the War on Terror: Life Struggles, Liberal Modernity, and the Defense of Logistical Societies* (Manchester: Manchester University Press, 2007); Michael Dillon, *The Liberal Way of War: The Martial Face of Global Politics* (London: Routledge, 2007); Derek Gregory and Allan Pred (eds) *Violent Geographies: Fear, Terror, and Political Violence* (New York: Routledge, 2007); and Michael Hardt and Antonio Negri, *Multitude: War and Democracy in the Age of Empire* (New York: Penguin, 2004).

5 Among the studies that have sought to rediscover Schmitt's political theory of sovereign exceptionalism, see Giorgio Agamben, *State of Exception* (Chicago: University of Chicago Press, 2005); Aihwa Ong, *Neoliberalism as Exception: Mutations in Citizenship and Sovereignty* (Durham, NC: Duke University Press, 2006); Jean-Claude Paye, *Global War on Liberty* (New York: Telos Press, 2007); Derek Gregory, "Vanishing Points: Law, Violence, and Exception in the Global War Prison," in Gregory and Pred (eds) *Violent Geographies*, pp. 205–36; Anna Secor, "An Unrecognizable Condition Has Arrived: Law, Violence, and the State of Exception in Turkey," in Gregory and Pred (eds), *Violent Geographies*, pp. 37–53; or Jef Huysmans, "International Politics of Exception: Competing Visions of International Political Order between Law and Politics," *Alternatives* 31 (2006), pp. 135–65.

6 The term "biopolitical sovereignty" is partially derived from Shapiro's own study of practices, regimes, and representations of cultural governance. See Michael Shapiro, *Methods and Nations: Cultural Governance and the Indigenous Subject* (New York: Routledge, 2004), pp. 180–82. See also François Debrix, *Tabloid Terror: War, Culture, and Geopolitics* (New York: Routledge, 2007), pp. 104–11.
7 Carl Schmitt, *Political Theology: Four Chapters on the Concept of Sovereignty* (Chicago: University of Chicago Press, 2005), p. 5.
8 See Carl Schmitt, "Ethic of State and Pluralistic State," in Chantal Mouffe (eds) *The Challenge of Carl Schmitt* (New York: Verso, 1999), pp. 195–208.
9 As Mouffe puts it: "what matters [to Schmitt] is the possibility of tracing a line of demarcation between those who belong to the demos—and therefore have equal rights—and those who, in the political domain, cannot have the same rights because they are not part of the demos." Mouffe adds: "[Schmitt's] main concern is not democratic participation but *political unity*." See Chantal Mouffe, "Carl Schmitt and the Paradox of Liberal Democracy," in Chantal Mouffe (eds) *The Challenge of Carl Schmitt*, pp. 41–42; author's emphasis.
10 See William Connolly, "The Complexities of Sovereignty," in Matthew Calarco and Steven DeCaroli (eds) *Giorgio Agamben: Sovereignty and Life* (Stanford, CA: Stanford University Press, 2007), pp. 24–25.
11 Giorgio Agamben, *Homo Sacer: Sovereign Power and Bare Life* (Stanford, CA: Stanford University Press, 1998), p. 11.
12 By contrast, in a recent essay, Pan argues that Schmitt's emphasis on the executive decision and the state of exception cannot be reduced to a "pure, mechanical violence" over individual life. Sovereign decisionism is rather to be understood "in terms of cultural assumptions about final goals of a society." For Pan, Schmitt's decisionism preserves the "ultimate values" that underpin the juridical order. Thus, Pan disputes Agamben's claim that the Schmittian state of exception is the zone of anomie, a space without law, through which the sovereign's authority and the juridical order are co-constituted, often with a potential for unmitigated violence. Pan minimizes the *political* decision that comes out of Schmitt's emphasis on the potentiality for violent confrontation between friend and enemy that originally gives meaning to the "ultimate values" of the polity. See David Pan, "Carl Schmitt on Culture and Violence in the Political Decision," *Telos* Vol. 142 (2008), pp. 49–72.
13 This is the case in Agamben's work, but also in writings by Michael Hardt and Antonio Negri. See Hardt and Negri, *Empire* (Cambridge: Harvard University Press, 2001); and Hardt and Negri, *Multitude*.
14 Achille Mbembe, "Necropolitics," *Public Culture* Vol. 15 (2003), p. 22.
15 Astrid Deuber-Mankowsky, "Nothing Is Political, Everything Can Be Politicized: On the Concept of the Political in Michel Foucault and Carl Schmitt," *Telos* Vol. 142 (2008), pp. 154–55.
16 Mouffe, "Carl Schmitt and the Paradox of Liberal Democracy," p. 42.
17 See, in particular, Michel Foucault, *Discipline and Punish: The Birth of the Prison* (New York: Vintage Books, 1979).
18 Agamben, *Homo Sacer*, p. 9.
19 For another problematization of Agamben's treatment of Foucault's biopower, see Paul Patton, "Agamben and Foucault on Biopower and Biopolitics," in Matthew Calarco and Steven DeCaroli (eds) *Giorgio Agamben: Sovereignty and Life* (Stanford, CA: Stanford University Press, 2007), pp. 203–18. Patton suggests that Agamben's interpretation of Foucault is driven by his desire to focus on the "why" of sovereign power whereas Foucault's concern is with the "how." See Patton, ibid., p. 212.
20 Connolly, "The Complexities of Sovereignty," pp. 30–31.

21 Paul Hirst, "Carl Schmitt's Decisionism," in Mouffe (ed.), *The Challenge of Carl Schmitt* (New York: Verso, 1999), p. 9.

22 Slavoj Žižek, "Carl Schmitt and the Age of Post-Politics," in Mouffe (eds) *The Challenge of Carl Schmitt*, p. 18.

23 This is not to say that we believe that Schmitt must be interpreted as a proponent of absolute slaughter, the ultimate war experience, or even fascistic politics, as Jeffrey Herf or Richard Wolin (among others) have argued. See Jeffrey Herf, *Reactionary Modernism: Technology, Culture, and Politics in Weimar and the Third Reich* (Cambridge: Cambridge University Press, 1984); and Richard Wolin, "Carl Schmitt: The Conservative Revolutionary Habitus and the Aesthetics of Horror," *Political Theory* Vol. 20 (1992), pp. 424–47. But it is also not enough to claim, as Joseph Bendersky all too hastily does, that "one will not find in Schmitt's writings any ideas, references, or resonances identifying him with the 'war experience' … [and that] it is absurd … to pursue this line of argument further." See Joseph Bendersky, "Carl Schmitt and the Conservative Revolution," *Telos* Vol. 72 (1987), pp. 28–29. While it is far too convenient to peg Schmitt as an apologist of state absolutism leading to military conquest and destruction (and possibly political nihilism), it is equally misleading to claim that Schmitt's writings on the sovereign exception are incapable of opening up the political onto instances of abyssal violence, whether Schmitt intended it to be the case or not. The point is not to identify Schmitt as a supporter of this or that political or ideological cause. This line of argumentation, we believe, has run its course. As we suggest at the end of this essay, the aporias, paradoxes, and inconsistencies in Schmitt's work are too important to brush aside to the benefit of one clear and prevailing interpretation (whether it is left-leaning or, on the contrary, right-wing). For a more balanced yet intricate examination of the tension between conflict and politics in Schmitt, see William Rasch, "Conflict as a Vocation: Carl Schmitt and the Possibility of Politics," *Theory, Culture, and Society* Vol. 17 (2000), pp. 1–32.

24 Andreas Kalyvas, "From the Act to the Decision: Hannah Arendt and the Question of Decisionism," *Political Theory* Vol. 32 (2004), p. 329.

25 Ibid., p. 329.

26 Lisa Disch, *Hannah Arendt and the Limits of Philosophy* (Ithaca, NY: Cornell University Press, 1994), p. 71.

27 Webster's states that "agony" is "a violent display of feelings or emotions." See *Webster's Dictionary* (New York: Merriam-Webster, 1953).

28 This meaning is well rendered in Luke 22:44: "Being in agony, he prayed more earnestly."

29 Arendt, *The Human Condition* (Chicago: University of Chicago Press, 1998), second edition, p. 205.

30 As we indicated above, while the adjective "agonistic" is more typically employed to refer to Arendt's work, we choose to use the term "agonal" to demarcate our purposes in this essay from Arendt's politics. For an excellent study of the stakes involved in the use of an Arendtian agonistic framework, see Andrew Schaap, "Political Theory and the Agony of Politics," *Political Studies Review* Vol. 5 (2007), pp. 56–74.

31 Arendt, *The Human Condition*, p. 205.

32 Hannah Arendt, *The Promise of Politics* (New York: Schoken, 2005), p. 45.

33 Ibid., p. 123.

34 Seyla Benhabib, *The Reluctant Modernism of Hannah Arendt* (Lanham: Rowman and Littlefield, 2003), p. 126.

35 Ibid., p. 129.

36 Bonnie Honig, *Political Theory and the Displacement of Politics* (Ithaca, NY: Cornell University Press, 1993), p. 89.

37 Dana Villa, *Arendt and Heidegger: The Fate of the Political* (Princeton, NJ: Princeton University Press, 1996), p. 91.
38 See, in particular, Jürgen Habermas, "Hannah Arendt's Communications Concept of Power," in Steven Lukes (ed.), *Power: Readings in Social and Political Theory* (New York: New York University Press, 1986), pp. 75–93.
39 Alain Joxe, *Empire of Disorder* (Los Angeles: Semiotext(e), 2002), p. 167; our emphases and inserts.
40 Hardt and Negri, *Multitude*, p. 12.
41 Ibid., p. 31; our inserts.
42 See Gilles Deleuze and Felix Guattari, *A Thousand Plateaus: Capitalism and Schizophrenia* (Minneapolis: University of Minnesota Press, 1987), pp. 474–500.
43 Andrew Neal, "Cutting Off the King's Head: Foucault's Society Must Be Defended and the Problem of Sovereignty," *Alternatives* Vol. 29 (2004), p. 378.
44 Ibid., p. 379.
45 Michel Foucault, *"Society Must Be Defende": Lectures at the Collège de France, 1975–76* (New York: Picador, 2003), p. 50.
46 Ibid., pp. 207–08.
47 Mbembe, "Necropolitics," p. 39.
48 Ibid., p. 30.
49 Andrew Bacevich, *The New American Militarism: How Americans Are Seduced by War* (New York: Oxford University Press, 2005), p. 15.
50 For a more detailed analysis of these narratives of war and their effects on public culture, see Debrix, *Tabloid Terror*.
51 Larry N. George, "American Insecurities and the Ontopolitics of US Pharmacotic Wars," in François Debrix and Mark J. Lacy (eds), *The Geopolitics of American Insecurity: Terror, Power, and Foreign Policy* (London: Routledge, 2009), p. 45.
52 Thomas Barnett, *The Pentagon's New Map: War and Peace in the Twenty-First Century* (New York: Putnam's Sons, 2004). See also his more recent *Blueprint for Action: A Future Worth Creating* (New York: Putnam's Sons, 2005). For an excellent critique of Barnett, see Simon Dalby, "The Pentagon's New Imperial Cartography," in Gregory and Pred (eds) *Violent Geographies*, pp. 295–308.
53 See P.W. Singer, *Wired for War: The Robotics Revolution and Conflict in the 21st Century* (New York: Penguin, 2009), p. 34.
54 As Wolfe puts it. See Cary Wolfe, *What Is Posthumanism?* (Minneapolis: University of Minnesota Press, 2010), p. xv.
55 Singer, *Wired for War*, p. 397.
56 An example of dehumanization of the kill by way of drones is the incident that took place in the border area between Afghanistan and Pakistan on February 21, 2010 when an unmanned US drone intended to track down "insurgents" attacked a civilian convoy, killing (in fact, butchering) mostly Afghan women and children. The initial drone spotting and attack was further followed by US helicopter fire as the helicopter crew was responding to the same flawed initial information captured and relayed by the drone. As a result, the innocent civilians were not merely killed, but in fact killed over and over, their bodies being made the object of multiple rounds of destructive force. See "Drone Crew Blamed in Afghan Civilian Deaths," *USA Today*, May 29, 2010, available at www.usatoday.com/news/world/afghanistan/2010-05-29-afghanistan-civilian-deaths_N.htm (accessed on April 18, 2011). Former British Senior Law Lord, Lord Bingham, famously stated in 2009 that unmanned drones could be "so cruel as to be beyond the pale of human tolerance." See Murray Waldrop, "Unmanned Drones Could Be Banned, Says Senior Judge," *The Daily Telegraph*, July 6, 2009, available at www.telegraph.co.uk/news/newstopics/politics/defence/5755446/Unmanned-drones-could-be-banned-says-senior-judge.html (accessed on April 18, 2011).
57 Neal, "Cutting Off the King's Head," p. 397.

58 See Primo Levi, *Si c'est un Homme* (Paris: Pocket Juliard, 2008), p. 310.
59 Schmitt, *The Concept of the Political*, p. 33.
60 Karl Löwith, "The Occasional Decisionism of Carl Schmitt," in Richard Wolin (ed.) *Martin Heidegger and European Nihilism* (New York: Columbia University Press, 1995), p. 147.
61 Ibid., p. 148.
62 Ibid., p. 148.
63 Carl Schmitt, *The Nomos of the Earth in the International Law of the Jus Publicum Europaeum* (New York: Telos Press, 2003).
64 Carl Schmitt, *Theory of the Partisan: Intermediate Commentary on the Concept of the Political* (New York: Telos Press, 2007), pp. 74–76.
65 Jacques Derrida, *Politics of Friendship* (New York: Verso, 1997), p. 86.

2 Nothing to fear but fear itself: governmentality and the reproduction of terror

1 Zygmunt Bauman, *Liquid Fear* (Cambridge: Polity, 2006), p. 96.
2 See Adriana Cavarero, *Horrorism: Naming Contemporary Violence* (New York: Columbia University Press, 2009).
3 George W. Bush, "President Discusses War on Terror at National Endowment for Democracy," available at www.whitehouse.gov/news/releases/2005/10/20051006-3.html (accessed January 8, 2008).
4 Jean-Claude Paye, *Global War on Liberty* (New York: Telos, 2007).
5 See Judith Butler, *Precarious Life: The Powers of Mourning and Violence* (London: Verso, 2004); Wendy Brown, *Regulating Aversion: Tolerance in the Age of Identity and Empire* (Princeton, NJ: Princeton University Press, 2006); and Paye, *Global War on Liberty.*
6 See Michael Hardt and Antonio Negri, *Multitude* (New York: Penguin, 2004).
7 See Michel Foucault, *The History of Sexuality: An Introduction*. Vol. 1 (New York: Vintage Books, 1990); Michel Foucault, *"Society Must Be Defended": Lectures at the Collège de France, 1975–76* (New York: Picador, 2003); and Michel Foucault, *Security, Territory, Population: Lectures at the Collège de France, 1977–78* (New York: Palgrave, 2007).
8 See Butler, *Precarious Life*, p. 65; and Andrew Neal, "Goodbye War on Terror? Foucault and Butler on Discourses of Law, War, and Exceptionalism," in Mick Dillon and Andrew Neal (eds), *Foucault on Politics, Security and War* (Basingstoke: Palgrave, 2008), pp. 43–64.
9 Corey Robin, *Fear: The History of a Political Idea* (Oxford: Oxford University Press, 2004), p. 31.
10 On this matter, see, for example, Leo Strauss, *The Political Philosophy of Hobbes: Its Basis and Its Genesis* (Chicago: University of Chicago Press, 1984).
11 See Thomas Hobbes, *The Elements of Law, Natural and Politic: Part I, Human Nature; Part II, De Corpore Politico; with Three Lives* (Oxford: Oxford University Press,1999), p. 22.
12 Ibid., p. 70.
13 Ibid., pp. 78–79.
14 Thomas Hobbes, *On the Citizen* (Cambridge: Cambridge University Press, 1998), p. 26.
15 Thomas Hobbes, *Leviathan* (New York: Penguin, 1985), p. 186.
16 Hobbes, *On the Citizen*, pp. 78–79.
17 Hobbes, *Leviathan*, p. 186.
18 Ibid., p. 186.
19 See Jef Huysmans, *The Politics of Insecurity: Migration and Asylum in the EU* (London: Routledge, 2006), p. 53.

20 As Williams argues. See Michael C. Williams, *The Realist Tradition and the Limits of International Relations* (Cambridge: Cambridge University Press, 2005).
21 See Leo Strauss, "Notes on Carl Schmitt, *The Concept of the Political*," in Carl Schmitt, *The Concept of the Political* (Chicago: University of Chicago Press, 1996), pp. 89–92; and Tracy Strong, "Foreword: The Sovereign and the Exception," in Carl Schmitt, *Political Theology: Four Chapters on the Concept of Sovereignty* (Chicago: University of Chicago Press, 2005), p. xx.
22 See David Dyzenhaus, "Putting the State Back in Credit," in Chantal Mouffe (ed.), *The Challenge of Carl Schmitt* (London: Verso, 1999), pp. 75–91.
23 See Carl Schmitt, *The Concept of the Political* (Chicago: University of Chicago Press, 2007).
24 Carl Schmitt, *Political Theology: Four Chapters on the Concept of Sovereignty* (Chicago: University of Chicago Press, 2005), p. 5.
25 Schmitt, *The Concept of the Political*, pp. 26–27.
26 Ibid., pp. 19–20.
27 Ibid., p. 45.
28 Jacques Derrida, *Politics of Friendship* (London: Verso, 1997), pp. 112–37.
29 Slavoj Žižek, "Carl Schmitt in the Age of Post-Politics," in Chantal Mouffe (ed), *The Challenge of Carl Schmitt* (London: Verso, 1999), p. 18.
30 See Gershon Weiler, *From Absolutism to Totalitarianism: Carl Schmitt on Thomas Hobbes* (London: Hollowbrook, 1994), pp. 98–99.
31 See Hardt and Negri, *Multitude*, pp. 10–11.
32 Neil Smith, *The Endgame of Globalization* (London: Routledge, 2004); Herfried Münkler, *Empires* (Cambridge: Polity, 2007); Christopher Coker, *The Warrior Ethos: Military Culture and the War on Terror* (London: Routledge, 2007); and Paye, *Global War on Liberty.*
33 See Michel Foucault, *Discipline and Punish: The Birth of the Prison* (New York: Vintage Books, 1979).
34 Foucault, *"Society Must Be Defended"*, p. 29.
35 Ibid., pp. 207–8.
36 Ibid., p. 7.
37 Ibid., p. 9.
38 Ibid., p. 98.
39 See Foucault, *The History of Sexuality*, Vol. 1.
40 See Michel Foucault, *Madness and Civilization: A History of Insanity in the Age of Reason* (New York: Vintage Books, 1965); and Foucault, *Discipline and Punish.*
41 Foucault, *The History of Sexuality*, Vol. 1, pp. 136–38.
42 Foucault, *"Society Must Be Defended"*, p. 243.
43 Ibid., p. 244.
44 Ibid., p. 248.
45 Ibid., p. 246.
46 Ibid., p. 248.
47 Ibid., p. 249.
48 Ibid., p. 255.
49 See Mitchell Dean, "Foucault, Government, and the Enfolding of Authority," in A. Barry, T. Osborne, and N. Rose (eds), *Foucault and Political Reason: Liberalism, Neo-liberalism, and Rationalities of Government* (Chicago: University of Chicago Press, 1996), p. 211.
50 Foucault, *"Society Must Be Defended"*, p. 257.
51 Ibid., pp. 191–226.
52 Ibid., p. 257.
53 Brown, *Regulating Aversion*, p. 79.
54 Ibid., p. 80.

55 Michel Foucault, "Governmentality," in G. Burchell, C. Gordon, and P. Miller (eds), *The Foucault Effect: Studies in Governmentality* (Chicago: University of Chicago Press, 1991), p. 95.
56 Brown, *Regulating Aversion*, p. 81.
57 Nikolas Rose, "Governing 'Advanced' Liberal Democracies," in A. Barry, T. Osborne, and N. Rose (eds), *Foucault and Political Reason: Liberalism, Neo-liberalism, and Rationalities of Government* (Chicago: University of Chicago Press, 1996), p. 38.
58 See Giorgio Agamben, *What Is an Apparatus?* (Stanford, CA: Stanford University Press, 2009), p. 3.
59 Ibid., p. 14.
60 Ibid., p. 8.
61 See Gilles Deleuze, "What Is a *Dispositif*?" in Gilles Deleuze, *Two Regimes of Madness: Texts and Interviews, 1975–1995* (New York: Semiotext(e), 2006), p. 340.
62 As Bennett adds, dispositifs or assemblages "are not governed by any central head: no one materiality or type of material has sufficient competence to determine consistently the trajectory or impact of the group." Thus, "[t]he effects generated by an assemblage [dispositif] are, rather, emergent properties." See Jane Bennett, *Vibrant Matter: A Political Ecology of Things* (Durham, NC: Duke University Press, 2010), p. 24.
63 See Didier Bigo, "Globalized-in-Security: The Field and the Ban-Opticon," in N. Sakai and J. Solomon (eds), *Translation, Biopolitics, Colonial Difference* (Hong Kong: Hong Kong University Press, 2006), p. 123. On this topic, see also Claudio Minca, "Giorgio Agamben and the New Biopolitical Nomos," *Geografiska Annaler* Vol. 88B, No. 4 (2006), pp. 387–403; Didier Bigo, "Detention of Foreigners, States of Exception, and the Social Practices of Control of the Banopticon," in P. K. Rajaram and C. Grundy-Warr (eds), *Borderscapes: Hidden Geographies and Politics at Territory's Edge* (Minneapolis: University of Minnesota Press, 2007), pp. 3–34; and Nick Vaughan-Williams, "Borders, Territory, Law," *International Political Sociology* Vol. 2, No. 4 (2008), pp. 322–38.
64 See Foucault, *Discipline and Punish*, pp. 195–228.
65 Bigo, "Globalized-in-Security," p. 134.
66 We are referring to the killing by the British police of Brazilian electrician Jean-Charles de Menezes in the London Underground in July 2005 while on his way to work. For an excellent study of this shooting and its connection to biopower, see Nick Vaughan-Williams, "The Shooting of Jean Charles de Menezes: New Border Politics?" *Alternatives* Vol. 32, No. 2 (2007), pp. 177–95.
67 See Dean, "Foucault, Government, and the Enfolding of Authority," p. 211; our emphasis.
68 See Paye, *Global War on Liberty*, p. 245.
69 Roberto Esposito, *Bios: Biopolitics and Philosophy* (Minneapolis: University of Minnesota Press, 2008), p. 46.
70 See Cavarero, *Horrorism*, p. 7.
71 See René Girard, "Sacrifice as Sacral Violence and Substitution," in James G. Williams (ed.), *The Girard Reader* (New York: Crossroad, 1996), p. 77.
72 Ibid., p. 77.
73 See Giorgio Agamben, *Homo Sacer: Sovereign Power and Bare Life* (Stanford, CA: Stanford University Press, 1998).
74 See Michael Dillon, "Governing Terror: The State of Emergency of Biopolitical Emergence," *International Political Sociology* Vol. 1, No. 1 (2007), pp. 7–28.
75 Ibid., p. 8.
76 Ibid., p. 14.
77 Ibid., p. 8.

78 Cavarero, *Horrorism*, p. 5.
79 On this topic, see, for example, Geoffrey Whitehall, "The Aesthetic Emergency of the Avian Flu Affect," in François Debrix and Mark Lacy (eds), *The Geopolitics of American Insecurity* (London: Routledge, 2009), pp. 161–80.

3 The nomos of exception and the virtuality of geopolitical space

1 Carl Schmitt, *The Nomos of the Earth in the International Law of the Jus Publicum Europaeum* (New York: Telos, 2006), p. 42; our emphasis.
2 Giorgio Agamben, *Homo Sacer: Sovereign Power and Bare Life* (Stanford, CA: Stanford University Press, 1998), p. 19.
3 See, for example, Jean-François Kervégan, "Carl Schmitt and 'World Unity'," in Chantal Mouffe (ed.), *The Challenge of Carl Schmitt* (London: Verso, 1999), pp. 54–74; Marti Koskenniemi, "International Law as Political Theology: How to Read *Nomos der Erde*," *Constellations* Vol. 11, No. 4 (2004), pp. 492–511; Bruno Bosteels, "The Obscure Subject: Sovereignty and Geopolitics in Carl Schmitt's *The Nomos of the Earth*," *South Atlantic Quarterly* Vol. 104, No. 2 (2005), pp. 295–306; Louiza Odysseos, "Crossing the Line? Carl Schmitt on the 'Spaceless Universalism' of Cosmopolitanism and the War on Terror," in Louiza Odysseos and Fabio Petito (eds), *The International Political Thought of Carl Schmitt* (London: Routledge, 2007), pp. 124–43; and Fabio Petito, "Against World Unity: Carl Schmitt and the Western-centric and Liberal Global Order," in Louiza Odysseos and Fabio Petito (eds), *The International Political Thought of Carl Schmitt* (London: Routledge, 2007), pp. 166–84.
4 This is particularly the case for many of the essays in Stephen Legg (ed.), *Spatiality, Sovereignty, and Carl Schmitt: Geographies of the Nomos* (London: Routledge, 2011).
5 Agamben, *Homo Sacer*, p. 19.
6 Carl Schmitt, *Über die Drei Arten des Rechtswissenschaftlichen Denkens* (Berlin: Duncker & Humblot, 1993), p. 55; also quoted in Mika Ojakangas, "Carl Schmitt and the Sacred Origins of Law," *Telos* No. 157 (Summer 2009), p. 41.
7 Giorgio Agamben, *Remnants of Auschwitz* (New York: Zone Books, 2002), p. 47.
8 On the camp as the prototypical space of exception, see Claudio Minca, "The Return of the Camp," *Progress in Human Geography* Vol. 29, No. 4 (2005), pp. 405–12; Derek Gregory, "The Black Flag: Guantanamo Bay and the Space of Exception," *Geografiska Annaler* Vol. 88B, No. 4 (2006), pp. 405–27; and Bulent Diken and Carsten Bagge Laustsen, "The Camp," *Geografiska Annaler* Vol. 88B, No. 4 (2006), pp. 443–52.
9 For a recent example of the use of the virtual as false representation or as opposed to truth, see Andrew Wilson, *Virtual Politics: Faking Democracy in the Post-Soviet World* (New Haven, CT: Yale University Press, 2005).
10 Philippe Quéau, *Le Virtuel: Vertus et Vertiges* (Paris: Champ Vallon, 1994), p. 9; our translation.
11 Ibid., p. 10; our translation.
12 See Gilles Deleuze, *Bergsonism* (New York: Zone Books, 1991).
13 For Deleuze, this also entails reconsidering what the "possible" is and how it manifests itself. According to DeLanda, the Deleuzian possible is a figure of multiplicity that is best rendered by one of Deleuze's other complex notions, that of the "manifold." Deleuze's manifold is a "variable number of dimensions" that are marked by their relations of supplementarity, not complementarity or combination. Thus, the manifold (or Deleuze's multiplicity) never amounts or aims to return to a "defining unity." The manifold, then, DeLanda suggests, "becomes the space of possible states." This "space of the possible," or the multiple/manifold, is what is virtual, according to Deleuze, about actual or material systems or states

of being. See Manuel DeLanda, *Intensive Science and Virtual Philosophy* (London: Continuum, 2002), pp. 12–13.

14 Gilles Deleuze, *Difference and Repetition* (New York: Columbia University Press, 1994), p. 208.

15 Gilles Deleuze and Félix Guattari, *What Is Philosophy?* (New York: Columbia University Press, 1994), p. 156.

16 See Nick Vaughan-Williams, "Virtual Border (In)security," paper presented at the 2009 annual meeting of the International Studies Association, New York, February 15–18, 2009, p. 16. See also Nick Vaughan-Williams, "Borders, Territory, Law," *International Political Sociology* Vol. 2, No. 4 (2008), pp. 322–38; and Agamben, *Homo Sacer*, p. 111.

17 See Walter Benjamin, "On the Concept of History," in H. Eiland and M. Jennings (eds), *Walter Benjamin: Selected Writings, Volume 4* (Cambridge, MA: Harvard University Press, 2003), pp. 389–400.

18 See Michael Hardt and Antonio Negri, *Multitude* (New York: Penguin, 2004), p. 7.

19 See, for example, Didier Bigo, "Detention of Foreigners, States of Exception, and the Social Practices of Control of the Banopticon," in P.K. Rajaram and C. Grundy-Warr (eds), *Borderscapes: Hidden Geographies and Politics at Territory's Edge* (Minneapolis: University of Minnesota Press, 2007), pp. 3–34; Claudio Minca, "Giorgio Agamben and the New Biopolitical *Nomos*," *Geografiska Annaler* Vol. 88B, No. 4 (2006), pp. 387–403; Nick Vaughan-Williams, "The Shooting of Jean-Charles de Menezes: New Border Politics?", *Alternatives* Vol. 32, No. 2 (2007), pp. 177–95; Roxanne Doty, "States of Exception on the Mexico–US Border: Security, 'Decisions', and Civilian Border Patrols," *International Political Sociology* Vol. 1, No. 2 (2007), pp. 113–37; Jenny Edkins, "Whatever Politics," in Matthew Calarco and Steven DeCaroli (eds), *Giorgio Agamben: Sovereignty and Life* (Stanford, CA: Stanford University Press, 2007), pp. 70–91; and Gerry Kearns, "Bare Life, Political Violence, and the Territorial Structure of Britain and Ireland," in Derek Gregory and Alan Pred (eds), *Violent Geographies: Fear, Terror, and Political Violence* (New York: Routledge, 2007), pp. 7–35.

20 As the infamous shooting of Brazilian electrician Jean-Charles de Menezes revealed. See Vaughan-Williams, "The Shooting of Jean-Charles de Menezes," pp. 177–95; Minca, "Giorgio Agamben and the New Biopolitical *Nomos*," p. 387.

21 See Anna Secor, "'An Unrecognizable Condition Has Arrived': Law, Violence, and the State of Exception in Turkey," in Derek Gregory and Alan Pred (eds), *Violent Geographies: Fear, Terror, and Political Violence* (New York: Routledge, 2007), pp. 37–53.

22 See Mathew Coleman, "Deserting Sovereignty? The Securitization of Undocumented Migration in the United States," in François Debrix and Mark Lacy (eds), *The Geopolitics of American Insecurity: Terror, Power, and Foreign Policy* (New York: Routledge, 2009), pp. 107–25.

23 See Didier Bigo, "Globalized-in-Security: The Field and the Banopticon," in N. Sakai and J. Solomon (eds), *Translation, Biopolitics, Colonial Difference* (Hong Kong: Hong Kong University Press, 2006), p. 123.

24 Mitchell Dean, "*Nomos*: Word and Myth," in Louiza Odysseos and Fabio Petito (eds), *The International Political Thought of Carl Schmitt* (London: Routledge, 2007), p. 246.

25 Ibid., p. 246.

26 As Ojakangas puts it. See Ojakangas, "Carl Schmitt and the Sacred Origins of Law," p. 35.

27 Ibid., p. 35.

28 Schmitt, *The* Nomos *of the Earth*, p. 42.

29 Ibid., p. 48.

30 Ojakangas, "Carl Schmitt and the Sacred Origins of Law," p. 39.
31 Schmitt, *The Nomos of the Earth*, p. 44.
32 See Carl Schmitt, "*Nomos—Nahme*—Name," in C. Schmitt, *The Nomos of the Earth in the International Law of the Jus Publicum Europaeum* (New York: Telos, 2006), pp. 336–50.
33 Ibid., p. 340.
34 Dean, "*Nomos*: Word and Myth," p. 244.
35 Schmitt, "*Nomos—Nahme*—Name," p. 341.
36 Dean, "*Nomos*: Word and Myth," p. 244.
37 Deleuze and Guattari write that "even though the nomadic trajectory may follow trails and customary routes, it does not fulfill the function of the sedentary road, which is to parcel out a closed space to people … The nomadic trajectory does the opposite: it distributes people (or animals) in an open space, one that is indefinite and noncommunicating." See Gilles Deleuze and Felix Guattari, *A Thousand Plateaus: Capitalism and Schizophrenia* (Minneapolis: University of Minnesota Press, 1987), p. 380.
38 Ibid., p. 8.
39 Ojakangas, "Carl Schmitt and the Sacred Origins of Law," p. 42; our emphasis.
40 See Carl Schmitt, *Glossarium—Aufziechnunger des Jahre 1947–1951* (Berlin: Duncker & Humblot, 1988), p. 37; see also Mika Ojakangas, "A Terrifying World without an Exterior," in Louiza Odysseos and Fabio Petito (eds), *The International Political Thought of Carl Schmitt* (London: Routledge, 2007), p. 206.
41 Schmitt, *The Nomos of the Earth*, p. 78.
42 On this topic, see Odysseos, "Crossing the Line?" pp. 124–43.
43 Carl Schmitt, *The Concept of the Political* (Chicago: University of Chicago Press, 1996), p. 53.
44 Schmitt, *The Nomos of the Earth*, p. 246 and p. 270.
45 Ibid., p. 321.
46 On this topic, see Schmitt's reflections on the ills of the League of Nations in the 1920s and 1930s. Ibid., pp. 240–58.
47 Ojakangas, "A Terrifying World without an Exterior," p. 215.
48 Schmitt, "*Nomos—Nahme*—Name," p. 347; Schmitt's own emphases.
49 Ojakangas, "A Terrifying World without an Exterior," p. 215.
50 Schmitt, *Glossarium—Aufziechnunger des Jahre 1947–1951*, p. 37.
51 Dean, "*Nomos*: Word and Myth," p. 248.
52 Ibid., p. 248.
53 See, in particular, Derek Gregory, "Vanishing Points: Law, Violence, and Exception in the Global War Prison," in Derek Gregory and Alan Pred (eds), *Violent Geographies: Fear, Terror, and Political Violence* (New York: Routledge, 2007), pp. 205–36.
54 See Mick Dillon, "Correlating Sovereignty and Biopower," in Jenny Edkins, Veronique Pin-Fat, and Michael Shapiro, *Sovereign Lives: Power in Global Politics* (New York: Routledge, 2004), pp. 41–60; see also Mick Dillon, "Governing Terror: The State of Emergency of Biopolitical Emergence," *International Political Sociology* Vol. 1, No. 1 (2007), pp. 7–28.
55 Gregory, "The Black Flag," p. 407.
56 See Jean Baudrillard, *Simulations* (New York: Semiotext(e), 1983).
57 Ibid., p. 2.
58 Gregory, "The Black Flag," p. 405.
59 Or, to quote Agamben: "if the essence of the camp consists in the materialization of the state of exception … , then we must admit that *we find ourselves virtually* in the presence of a camp every time such a structure is created, … whatever its denomination and specific topography." See Agamben, *Homo Sacer*, p. 174; our emphasis.

60 As Baudrillard used to say about the Gulf War (that it "never took place"), precisely because its simulated scenarios already had been played out many times over, through war simulations and media images and games. See Jean Baudrillard, *The Gulf War Did Not Take Place* (Bloomington, IN: Indiana University Press, 1995).
61 On this topic, see François Debrix, *Re-Envisioning Peacekeeping: The United Nations and the Mobilization of Ideology* (Minneapolis: University of Minnesota Press, 1999).
62 This analysis also recalls Luke's reading of the turn to simulation and hyperreality in security studies and containment discourses in the aftermath of the Cold War. See Timothy W. Luke, "The Discipline of Security Studies and the Codes of Containment: Learning from Kuwait," *Alternatives* Vol. 16, No. 3 (1991), pp. 315–44.
63 See Vaughan-Williams, "Virtual Border (In)security," pp. 4–9.
64 Ibid., pp. 16–17.
65 Steven DeCaroli, "Boundary Stones: Giorgio Agamben and the Field of Sovereignty," in Matthew Calarco and Steven DeCaroli (eds), *Giorgio Agamben: Sovereignty and Life* (Stanford, CA: Stanford University Press, 2007), p. 47; our emphasis.
66 See Claudio Minca, "Agamben's Geographies of Modernity," *Political Geography* Vol. 26 (2007), p. 84.
67 Minca, "The Return of the Camp," p. 407.
68 Ibid., p. 407.
69 Ibid., p. 408.
70 Minca, "Giorgio Agamben and the New Biopolitical *Nomos*," p. 388.
71 Minca, "The Return of the Camp," p. 408.
72 Agamben, *Homo Sacer*, p. 9; Minca, "The Return of the Camp," p. 408.
73 Ibid., p. 409.
74 On the etymology of the term "virtual," see Quéau, *Le Virtuel*, p. 26.
75 Minca, "The Return of the Camp," p. 409.
76 Ibid., p. 410.
77 Michel Foucault, *"Society Must Be Defended": Lectures at the College de France, 1975–1976* (New York: Picador, 2003), pp. 239–40.
78 As Cavarero puts it. See Adriana Cavarero, *Horrorism: Naming Contemporary Violence* (New York: Columbia University Press, 2009), p. 43.
79 See Judith Butler, *Precarious Life: The Powers of Mourning and Violence* (New York: Verso, 2004), p. 29.
80 Cavarero, *Horrorism,* p. 30; our emphasis.

4 The horror of enmity: rethinking alterity in the age of Global War

1 Jacques Derrida, *The Politics of Friendship* (New York: Verso, 2005), p. 84.
2 Gil Anidjar, *The Jew, the Arab: A History of the Enemy* (Stanford, CA: Stanford University Press, 2003), pp. 168–70.
3 Michael Hardt and Antonio Negri, *Empire* (Cambridge, MA: Harvard University Press, 2000), pp. 210–11; Also quoted in Anidjar, *The Jew, the Arab,* p. 169.
4 Anidjar, *The Jew, the Arab*, p. 169.
5 See Adriana Cavarero, *Horrorism: Naming Contemporary Violence* (New York: Columbia, 2009), pp. 104–05.
6 Carl Schmitt, *The Concept of the Political* (Chicago: Chicago University Press, 2007), p. 27.
7 Mika Ojakangas, *A Philosophy of Concrete Life: Carl Schmitt and the Political Thought of Late Modernity* (New York: Peter Lang, 2006), p. 87.

8 Heinrich Meier, *The Lesson of Carl Schmitt: Four Chapters on the Distinction between Political Theology and Political Philosophy* (Chicago: University of Chicago Press, 1998), p. 44; our emphasis.
9 Carl Schmitt, *The Concept of the Political*, p. 67.
10 Derrida, *The Politics of Friendship*, p. 86.
11 Meier, *The Lesson of Carl Schmitt*, p. 45.
12 Ibid., p. 57
13 Susan Buck-Morss, *Dreamworld and Catastrophe: The Passing of Mass Utopia in East and West* (Cambridge, MA: MIT Press, 2000), p. 32.
14 Ibid., Part I.
15 Ibid., p. 12.
16 Schmitt, *The Concept of the Political*, p. 28.
17 Ibid., p. 63.
18 Buck-Morss, *Dreamworld and Catastrophe*, p. 33.
19 Ibid., pp. 12–13.
20 Susan Buck-Morss, "Sovereign Right and the Global Left," *Cultural Critique* No. 69 (2008), p. 145.
21 Ibid., p. 145.
22 Buck-Morss, *Dreamworld and Catastrophe*, p. 13.
23 Ibid., p. 13.
24 Derrida, *The Politics of Friendship*, p. 114.
25 Schmitt, *The Concept of the Political*, pp. 27–28; our emphasis.
26 Derrida, *The Politics of Friendship*, pp. 87–88; quoted in Nick Mansfield, *Theorizing War: From Hobbes to Badiou* (New York: Palgrave Macmillan, 2008), p. 106.
27 Schmitt, *The Concept of the Political*, p. 29.
28 Derrida, *The Politics of Friendship*, p. 89.
29 Ibid., p. 89; our emphases.
30 Mansfield, *Theorizing War*, p. 108.
31 This reading also renders the conventional interpretation that Schmitt's thought is driven by the need to "bracket war" between legitimate public enemies problematic. A similar argument has been made by Barder. See Alexander D. Barder, "Lessons from the Grand Inquisitor: Carl Schmitt and the Providential Enemy," *Theory & Event* Vol. 12, No. 3 (2009), available at http://muse.jhu.edu/login?uri=/journals/theory_and_event/v012/12.3.barder.html (accessed September 13, 2010).
32 Derrida, *The Politics of Friendship*, p. 91.
33 Schmitt, *The Concept of the Political*, p. 30; our inserts.
34 Something that Foucault started to discuss as early as in *Madness and Civilization*. See Michel Foucault, *Madness and Civilization: A History of Insanity in the Age of Reason* (New York: Vintage Books, 1965).
35 Michel Foucault, *Security, Territory, Population: Lectures at the Collège de France 1977–78* (New York: Palgrave, 2009), p. 57.
36 Foucault, *Discipline and Punish: The Birth of the Prison* (New York: Vintage Books, 1979), p. 198.
37 Jeffrey T. Nealon, *Foucault Beyond Foucault: Power and Its Intensifications since 1984* (Stanford, CA: Stanford University Press, 2008), p. 51; our emphasis.
38 Foucault, *"Society Must Be Defended": Lectures at the Collège De France, 1975–76*, (New York: Picador, 2003), p. 258.
39 See Jacques Derrida, *Rogues: Two Essays on Reason* (Stanford, CA: Stanford University Press, 2005).
40 Michael Dillon and Julian Reid, *The Liberal Way of War: Killing to Make Life Live* (London: Routledge, 2009), p. 142.
41 Ibid., p. 142.
42 Foucault, *"Society Must Be Defended"*, p. 259.

43 Roberto Esposito, *Bíos: Biopolitics and Philosophy* (Minneapolis: University of Minnesota Press, 2008), p. 115.
44 Ibid., p. 115.
45 Ibid., p. 138.
46 Larry N. George, "American Insecurities and the Ontopolitics of Us Pharmacotic Wars," in François Debrix and Mark J. Lacy (eds), *The Geopolitics of American Insecurity: Terror, Power, and Foreign Policy* (New York: Routledge, 2009), pp. 34–53.
47 Anne McClintock, "Paranoid Empire: Specters from Guantánamo and Abu Ghraib," *Small Axe* Vol. 13, No. 1 (2009), p. 50.
48 Ibid., p. 57; author's emphasis.
49 Ibid., p. 58.
50 Carlo Galli, "On War and the Enemy," *CR: The New Centennial Review* Vol. 9, No. 2 (2009), p. 215.
51 Ibid., p. 215; author's emphasis.
52 Ibid., p. 216.
53 Ibid., p. 217.
54 Ibid., p. 217.
55 Gilles Deleuze and Felix Guattari, *A Thousand Plateaus: Capitalism and Schizophrenia* (Minneapolis: University of Minnesota Press, 1987), p. 422.
56 Brian Massumi, "Everywhere You Want to Be: Introduction to Fear," in Brian Massumi (ed.), *The Politics of Fear* (Minneapolis: University of Minnesota Press, 1993), p. 11.
57 Ibid., p. 11.
58 Ibid., p. 11; author's emphasis.
59 See, in particular, "Afghan Killings, Body Parts at Center of Inquiry," Msnbc.com, September 27, 2010, available at www.msnbc.msn.com/id/39380805/from/toolbar# (accessed on October 27, 2010). See also Hal Bernton, "Fast-Paced Probe Laid Foundation for Murder Charges against Lewis-McChord Soldiers," *The Seattle Times*, October 27, 2010, available at seattletimes.nwsource.com/cgi-bin/PrintStory.pl?document_id=2013265498 (accessed on October 27, 2010).
60 One article mentions that the soldiers "mutilated Afghan corpses and even collected fingers and other body parts, and that some [soldiers] posed for photos with Afghan corpses." That same article referred to the case as the "grimmest investigation of alleged atrocities by US military personnel" during the war in Afghanistan. See "Afghan Killings, Body Parts at Center of Inquiry," available at www.msnbc.msn.com/id/39380805/from/toolbar# (accessed on April 20, 2011).
61 Adriana Cavarero, *Horrorism: Naming Contemporary Violence* (New York: Columbia University Press, 2009), p. 8.
62 See Grégoire Chamayou, *Les Chasses à l'Homme* (*Manhunts*) (Paris: La Fabrique, 2010).
63 Ibid., p. 26.
64 See Michel Foucault, *Security, Territory, Population: Lectures at the Collège de France, 1977–78* (New York: Palgrave, 2007), pp. 123–30.
65 The etymological roots of the term "cynegetic" are the Latin Cynegiticus and the Greek Kynegetikos that mean hunting, particularly the idea of a chase led with dogs. "The Cynegeticus" was also the title of Xenophon's text typically translated as "On Hunting." See A.A. Phillips and M.M. Willcock (eds), *Xenophon and Arrian on Hunting* (New York: Aris and Phillips, 1999).
66 Chamayou, *Les Chasses à l'Homme*, p. 29.
67 Ibid., pp. 28–29.
68 Ibid., p. 28.
69 As Cavarero puts it. See Cavarero, *Horrorrism*, p. 8.
70 Chamayou, *Les Chasses à l'Homme*, p. 27.

71 Julia Kristeva, *Powers of Horror: An Essay on Abjection* (New York: Columbia University Press, 1982), p. 1.
72 Ibid., p. 1.
73 Ibid., p. 2.
74 See, in particular. Jacques Derrida, *Rogues: Two Essays on Reason* (Stanford, CA: Stanford University Press, 2005), and Jean-Luc Nancy, *The Experience of Freedom* (Stanford, CA: Stanford University Press, 1993).
75 This apt definition of ontopolitics is provided by George. See George, "American Insecurities and the Ontopolitics of US Pharmacotic Wars," p. 37.
76 See Michael Hardt and Antonio Negri, *Multitude* (New York: Penguin, 2004), p. 192.
77 Achille Mbembe, "On Politics as a Form of Expenditure," in John L. Comaroff and Jean Comaroff (eds), *Law and Disorder in the Postcolony* (Chicago: University of Chicago Press, 2006), p. 306.
78 Ibid., p. 305.
79 Ibid., p. 305.
80 Or, perhaps, as Chamayou might put it, Mbembe's idea of a new horrific form of "African governmentality" is more akin to the notion of "cynegetic" power or force—the power of the hunt, once again—than with pastoral power.
81 Georges Bataille, *The Accursed Share: An Essay on General Economy, Volume I* (New York: Zone Books, 1991), p. 21.
82 See, for example, René Girard, *Violence and the Sacred* (Baltimore, MD: Johns Hopkins University Press, 1977).
83 Cavarero too finds an important connection between Bataille's thought and her perspective on horror. She writes: "[f]or Bataille, my opening myself to 'communion' [i.e., for Bataille, the inevitable connection of one's body to others, be they friends or enemies], as well as obviously implicating the other, implies above all that I destroy 'the integrity of being, in myself and in the other'." See Cavarero, *Horrorism*, p. 58; our inserts and emphasis.

Conclusion: facing horrific violence

1 Michel Foucault, *Discipline and Punish: The Birth of the Prison* (New York: Vintage Books, 1979), p. 33.
2 Foucault, *Discipline and Punish*, pp. 3–5.
3 Ibid., p. 3.
4 Ibid., p. 3.
5 Ibid., p. 3.
6 Ibid., p. 5.
7 Ibid., p. 8.
8 See Alma Guillermoprieto, "The Murders of Mexico," *The New York Review of Books*, October 28, 2010, available at www.nybooks.com/archives/2010/oct/28/murderers-mexico/ (accessed on November 16, 2010).
9 Ed Vulliamy "The Terror," *Vanityfair*, October 21, 2010, available at *http://www.vanityfair.com/politics/features/2010/10/drug-wars-in-mexico-201010?printable=true* (accessed on November 16, 2010). See also Ed Vulliamy, *Amexica: War along the Borderline* (New York: Farrar, Straus and Giroux, 2010).
10 Vulliamy, "The Terror," no page given.
11 Ibid., no page given.
12 Adriana Cavarero, *Horrorism: Naming Contemporary Violence* (New York: Columbia University Press, 2009), pp. 4–9.
13 See Grégoire Chamayou, *Les Chasses à l'Homme* (Paris: La Fabrique, 2010).
14 Foucault, *Discipline and Punish*, p. 32.
15 See Carlo Galli, *Political Spaces and Global War* (Minneapolis: University of Minnesota Press, 2010).

16 Vulliamy, "The Terror," no page given.
17 Foucault, *Discipline and Punish*, p. 33.
18 Ibid., p. 33.
19 Ibid., p. 33.
20 Foucault, citing Ollyffe, *An Essay to Prevent Capital Crimes* (1731), ibid., p. 34.
21 Ibid., p. 9.
22 Ibid., p. 34.
23 Ibid., p. 34.
24 Ibid., p. 34.
25 Chamayou, *Les Chasses à l'Homme*, p. 27.
26 Vulliamy, "The Terror," no page given.
27 See Gilles Deleuze and Felix Guattari, *A Thousand Plateaus: Capitalism and Schizophrenia* (Minneapolis: University of Minnesota Press, 1987), pp. 351–423.
28 As Dyer-Witherford and de Peuter suggest, the sovereign majesty of Deleuze and Guattari's war machine is premised upon its nomadic qualities of mobility and subversiveness, not on fixity, centrality, or hegemony. See Nick Dyer-Witherford and Greig de Peuter, *Games of Empire: Global Capitalism and Video Games* (Minneapolis: University of Minnesota Press, 2009), pp. 84–85.
29 Slavoj Žižek, "Carl Schmitt in the Age of Post-Politics," in Chantal Mouffe (ed.), *The Challenge of Carl Schmitt* (London: Verso, 1999), p. 18.
30 On this issue, see, for example, Juan Carlos Garzon, *Mafia & Co: The Criminal Networks in Mexico, Brazil, and Colombia* (Washington, DC: Woodrow Wilson International Center for Scholars, first published in Spanish in 2008; no year given for English translation), available at www.wilsoncenter.org (accessed on December 20, 2010). See also "Mexico's Drug War: Falling Kingpins, Rising Violence", *The Economist*, December 16, 2010, available at www.economist.com/node/17733279?story_id=17733279&CFID=157801906&CFTOKEN=30578643 (on December 27, 2010).
31 Foucault, *Discipline and Punish*, pp. 8–9.
32 Foucault writes that the public execution became "a hearth in which violence bursts again into flame." See ibid., p. 9.
33 Deleuze and Guattari, *A Thousand Plateaus*, p. 355.
34 This is, of course, not to excuse or even justify the horrendous violence onto the flesh of humanity carried out by agonal warriors, whether they are vengeful masses in eighteenth century France, US soldiers "gone rogue" in Afghanistan, or drug war-related groups like Los Zetas or La Familia in Mexico. The point is rather to intimate that the spectacle of horror that insists on targeting the body beyond the threshold of life and death may not be so easily distinguishable from or foreign to the political violence and terror of modern power, modern sovereignty, and the modern state (in its many forms).
35 Cavarero, *Horrorism*, p. 43. Thus, for example, there is a certain coincidence between biopolitical analyses of the Holocaust and the idea that the Nazi death machine was a faceless industrial endeavor reflecting the supremacy of technological means of production and thinking. By contrast, the historian Timothy Snyder has recently argued that this "image is too simple, too clean." Snyder believes that a more accurate understanding of the horror of Nazism, along with that of Stalinism in the 1930s and 1940s, entails recognizing that mass shootings and starvation first took place, and that only subsequently gassing occurred in places like Poland, Ukraine, Belorussia, and the Baltic States, or what he renames the "Bloodlands." See Timothy Snyder, *Bloodlands: Europe between Hitler and Stalin* (New York: Basic Books, 2010), p. xiv. For a discussion of the relationship between technology and radical evil, see Slavoj Žižek, *In Defense of Lost Causes* (New York: Verso, 2009), pp. 142–47.

36 Cavarero, *Horrorism*, p. 43; Hannah Arendt, *The Origins of Totalitarianism* (New York: Harcourt, 1994), p. 443.
37 Primo Levi, *The Drowned and the Saved* (New York: Vintage International, 1989), Chapter 5.
38 See Alain Resnais' film *Nuit et Brouillard* [Night and Fog] (1955).
39 Levi, *The Drowned and the Saved*, p. 120.
40 Arendt, *The Origins of Totalitarianism*, p. 443.
41 See, in particular, Michael J. Shapiro, *Methods and Nations: Cultural Governance and the Indigenous Subject* (New York: Routledge, 2003), p. 180; François Debrix, *Tabloid Terror: War, Culture, and Geopolitics* (New York: Routledge, 2008), pp. 105–07.
42 See Gilles Deleuze and Félix Guattari, *Anti-Oedipus: Capitalism and Schizophrenia* (Minneapolis: University of Minnesota Press, 1983), pp. 9–16.
43 Jacques Derrida, "Deconstruction in America: An Interview with Jacques Derrida," *Critical Exchange*, Vol. 17 (1985), pp. 1–32.
44 Bonnie Honig, *Emergency Politics: Paradox, Law, Democracy* (Princeton, NJ: Princeton University Press, 2009).
45 Ibid., p. 10.
46 Ibid., p. 10.
47 Ibid., p. 10.
48 Ibid., p. 10.
49 Didier Fassin, "Ethics of Survival: A Democratic Approach to the Politics of Life," *Humanity: An International Journal of Human Rights, Humanitarianism, and Development*, Vol. 1, No. 1 (2010), p. 83. Fassin also seeks to demonstrate in this essay (although unsuccessfully, we believe) that Derrida's so-called ethic of survival can provide a way out of the biopolitical logic advanced by Foucault, Agamben, and others. On this issue, see also Didier Fassin, "Another Politics of Life is Possible," *Theory, Culture & Society*, Vol. 26, No. 5 (2009), pp. 44–60.
50 Fassin, "Ethics of Survival," p. 83.
51 On the matter of the logic of the chance event as a gift or a dividend that cannot be earned, see Jacques Derrida, *Given Time: I. Counterfeit Money* (Chicago: University of Chicago Press, 1992). See also Jacques Derrida, *Rogues: Two Essays on Reason* (Stanford, CA: Stanford University Press, 2005), pp. 84–85.
52 Honig, *Emergency Politics*, p. 10.
53 Again, a similar critique can be levied at Fassin's argument about a new politics of life and his use of Derrida's thought, as we hinted at above.
54 See Rosi Braidotti, "The Politics of 'Life Itself' and New Ways of Dying," in Diana Coole and Samantha Frost (eds), *New Materialisms: Ontology, Agency, and Politics* (Durham, NC: Duke University Press, 2010), p. 201.
55 See Judith Butler, *Precarious Life: The Powers of Mourning and Violence* (New York: Verso, 2006); see more recently Judith Butler, *Frames of War: When Is Life Grievable?* (New York: Verso, 2009).
56 Butler, *Frames of War*, p. 33.
57 Ibid., p. 52.
58 Judith Butler, *Bodies that Matter: On the Discursive Limits of Sex* (New York: Routledge, 1993).
59 Butler, *Frames of War*, pp. 52–53.
60 Ibid., p. 54.
61 Or, perhaps, what Agier has recently reformulated as "humanity as an identity." See Michel Agier, "Humanity as an Identity and Its Political Effects," *Humanity: An International Journal of Human Rights, Humanitarianism, and Development* Vol. 1, No. 1 (2010), pp. 29–31.
62 Butler, *Frames of War*, p. 53.

63 As we showed in the "Introduction" to this volume. See also Butler, *Frames of War*, pp. 66–74.
64 Ibid., p. 53.
65 Ibid., p. 62.
66 Ibid., p. 55.
67 Ibid., p. 59.
68 Ibid., p. 15.
69 Ibid., p. 60. Butler's quote here borrows some of the same words found in Guantanamo's written fragments.
70 This point indirectly recalls Weisband's suggestion that responses to atrocities in the form of forgiveness or bereavement often "fail to accommodate the depths and complexities of grief and mourning." For Weisband, a reconsideration of the bereavement and grief as political forms is in order. See Edward Weisband, "On the Aporetic Borderlines of Forgiveness: Bereavement as a Political Form," *Alternatives* Vol. 34, No. 4 (2009), p. 360.
71 Hannah. Arendt, *The Life of the Mind* (New York: Harcourt, 1981), p. 78.
72 Michael J. Shapiro, *Cinematic Geopolitics* (London: Routledge, 2009), p. 5.
73 Cora Diamond, "The Difficulty of Reality and the Difficulty of Philosophy," in Stanley Cavell, Cora Diamond, John McDowell, Ian Hacking, and Cary Wolfe (eds), *Philosophy and Animal Life* (New York: Columbia University Press, 2009), pp. 43–90.
74 Ibid., p. 58.
75 Ibid., p. 58. See also J. M. Coetzee, *The Lives of Animals* (Princeton, NJ: Princeton University Press, 1999).
76 Diamond, "The Difficulty of Reality and the Difficulty of Philosophy," p. 59.
77 Ibid., p. 59.
78 Ibid., p. 77.
79 We are most grateful to Scott Nelson for bringing this notion of an "ethic of inventorying the scattered" to our attention in his thoughtful response to our work.
80 In this way, the ethic of inventorying the scattered also tries to maintain the qualities of eventness and unpredictability to the moment or scene of horror. As Samuel Weber reminds us, the "unpredictable event" is anathema to systems of thought, meaning, and affect. The "unpredictable event" always "requires the suspension or alteration of the established grid." See Samuel Weber, *Targets of Opportunity: On the Militarization of Thinking* (New York: Fordham University Press, 2005), p. 5.
81 On the critical as well as disembodying and proliferating capacity of mimetic thinking, see Michael Taussig, *Mimesis and Alterity: A Particular History of the Senses* (New York: Routledge, 1993).
82 Thus, and for reasons we explained above, we disagree with Fassin's claim that a Derridean ethics of survival as *survivance* or as over-living is better suited to "open an ethical space for reflection and action." See Fassin, "Ethics of Survival," p. 93.
83 This is one of its main differences with Arendt's ethics of natality. See Hannah Arendt, *The Human Condition* (Chicago: University of Chicago Press, 1958).
84 For another complementary treatment of the political as an opening up of space, see Scott Nelson, *Sovereignty and the Limits of the Liberal Imagination* (London: Routledge, 2009).
85 As Cavarero stated: "[a]s violence spreads and assumes unheard-of forms, it becomes difficult to name in contemporary language." See Cavarero, *Horrorism*, p. 2.
86 On the importance of narrative coherence when it comes to "bearing witness," see James Dawes, *That the World May Know: Bearing Witness to Atrocity* (Cambridge, MA: Harvard University Press, 2007). On the need to find ways to

narrativize horror, see also Philip Gourevitch, *We Wish to Inform You That Tomorrow We Will Be Killed with Our Families: Stories from Rwanda* (New York: Picador, 1999).

87 Jean Baudrillard, *Carnival and Cannibal* (London: Seagull Books, 2010), pp. 22–23; author's emphasis.

88 See Georges Bataille, *The Accursed Share: An Essay on General Economy, Volume I* (New York: Zone Books, 1991). See also our analysis at the end of Chapter 4.

89 This statement is partially derived from Baudrillard's concept of a "theater of decomposition." See Baudrillard, *Carnival and Cannibal*, p. 12.

90 Ibid., pp. 27–28.

Bibliography

Agamben, Giorgio. *Homo Sacer: Sovereign Power and Bare Life* (Stanford, CA: Stanford University Press, 1998).

——. *Remnants of Auschwitz* (New York: Zone Books, 2002).

——. *State of Exception* (Chicago: University of Chicago Press, 2005).

——. *What is an Apparatus?* (Stanford, CA: Stanford University Press, 2009).

Agier, Michel. "Humanity as an Identity and Its Political Effects," *Humanity: An International Journal of Human Rights, Humanitarianism, and Development* Vol. 1, No. 1 (2010): 29–45.

Anidjar, Gil. *The Jew, the Arab: A History of the Enemy* (Stanford, CA: Stanford University Press, 2003).

Arendt, Hannah. *The Origins of Totalitarianism* (New York: Harcourt Brace Jovanovich, 1994).

——. *The Human Condition* (Chicago: University of Chicago Press, 1998).

——. *The Promise of Politics* (New York: Schoken, 2005).

Bacevich, Andrew. *The New American Militarism: How Americans Are Seduced by War* (New York: Oxford University Press, 2005).

Barder, Alexander D. "Lessons from the Grand Inquisitor: Carl Schmitt and the Providential Enemy," *Theory & Event* Vol. 12, No. 3 (2009), available at http://muse.jhu.edu/login?uri=/journals/theory_and_event/v012/12.3.barder.html (accessed on April 21, 2011).

Barnett, Thomas. *The Pentagon's New Map: War and Peace in the Twenty-First Century* (New York: Putnam's Sons, 2004).

Bataille, Georges. *The Accursed Share: An Essay on General Economy*, Volume I (New York: Zone Books, 1991).

Baudrillard, Jean. *Simulations* (New York: Semiotext(e), 1983).

——. *The Gulf War Did Not Take Place* (Bloomington, IN: Indiana University Press, 1995).

——. *Carnival and Cannibal* (London: Seagull Books, 2010).

Bauman, Zygmunt. *Liquid Fear* (Cambridge: Polity, 2006).

Benjamin, Walter. "On the Concept of History," in H. Eiland and M. Jennings (eds), *Walter Benjamin: Selected Writings, Volume 4* (Cambridge, MA: Harvard University Press, 2003).

Bendersky, Joseph. "Carl Schmitt and the Conservative Revolution," *Telos* Vol. 72 (1987): 27–42.

Benhabib, Seyla. *The Reluctant Modernism of Hannah Arendt* (Lanham, MD: Rowman and Littlefield, 2003).

Bennett, Jane. *Vibrant Matter: A Political Ecology of Things* (Durham, NC: Duke University Press, 2010).

Bigo, Didier. "Globalized-in-Security: The Field and the Ban-Opticon," in N. Sakai and J. Solomon (eds), *Translation, Biopolitics, Colonial Difference* (Hong Kong: Hong Kong University Press, 2006).

——. "Detention of Foreigners, States of Exception, and the Social Practices of Control of the Banopticon," in P. K. Rajaram and C. Grundy-Warr (eds), *Borderscapes: Hidden Geographies and Politics at Territory's Edge* (Minneapolis: University of Minnesota Press, 2007).

Bigo, Didier and Anastassia Tsoukala (eds). *Terror, Insecurity and Liberty: Illiberal Practices of Liberal Regimes after 9/11* (London: Routledge, 2009).

Bosteels, Bruno. "The Obscure Subject: Sovereignty and Geopolitics in Carl Schmitt's *The Nomos of the Earth*," *South Atlantic Quarterly* Vol. 104, No. 2 (2005): 295–306.

Braidotti, Rosi. "The Politics of 'Life' Itself and New Ways of Dying," in Diana Coole and Samantha Frost (eds), *New Materialisms: Ontology, Agency, and Politics* (Durham, NC: Duke University Press, 2010): 201–18.

Brown, Wendy. *Regulating Aversion: Tolerance in the Age of Identity and Empire* (Princeton, NJ: Princeton University Press, 2006).

Buck-Morss, Susan. *Dreamworld and Catastrophe: The Passing of Mass Utopia in East and West* (Cambridge, MA: MIT Press, 2000).

Burke, Anthony. *Beyond Security, Ethics and Violence: War against the Other* (London: Routledge, 2007).

Bush, George W. "President Discusses War on Terror at National Endowment for Democracy," 2005, available at www.whitehouse.gov/news/releases/2005/10/20051006-3.html (accessed on January 8, 2008).

Butler, Judith. *Bodies that Matter: On the Discursive Limits of Sex* (New York: Routledge, 1993).

——. *Precarious Life: The Powers of Mourning and Violence* (London: Verso, 2004).

——. *Frames of War: When is Life Grievable*? (New York: Verso, 2009).

Campbell, David. *National Deconstruction: Violence, Identity, and Justice in Bosnia* (Minneapolis: University of Minnesota Press, 1998).

——. "The Biopolitics of Security: Oil, Empire, and the Sports Utility Vehicle," *The American Quarterly* Vol. 57, No. 3 (2005): 943–72.

Cavarero, Adriana. *Horrorism: Naming Contemporary Violence* (New York: Columbia University Press, 2007).

Chamayou, Grégoire. *Les Chasses à l'Homme* (Paris: La Fabrique, 2010).

Chow, Rey. *The Age of the World Target: Self-Referentiality in War, Theory, and Comparative Work* (Durham, NC: Duke University Press, 2006).

Coetzee, J. M. *The Lives of Animals* (Princeton, NJ: Princeton University Press, 1999).

Coker, Christopher. *The Warrior Ethos: Military Culture and the War on Terror* (London: Routledge, 2007).

Coleman, Mathew. "Deserting Sovereignty? The Securitization of Undocumented Migration in the United States," in François Debrix and Mark Lacy (eds), *The Geopolitics of American Insecurity: Terror, Power, and Foreign Policy* (New York: Routledge, 2009).

Connolly, William. *The Ethos of Pluralization* (Minneapolis: University of Minnesota Press, 1995).

——. "The Complexities of Sovereignty," in Matthew Calarco and Steven DeCaroli (eds), *Giorgio Agamben: Sovereignty and Life* (Stanford, CA: Stanford University Press, 2007).

Croft, Stuart. *Culture, Crisis and America's War on Terror* (Cambridge: Cambridge University Press, 2006).

Dalby, Simon. "The Pentagon's New Imperial Cartography," in Derek Gregory and Allan Pred (eds), *Violent Geographies* (New York: Routledge, 2007).

Dauphinee, Elizabeth and Cristina Masters (eds). *The Logics of Biopower and the War on Terror* (London: Palgrave, 2006).

Dawes, James. *That the World May Know: Bearing Witness to Atrocity* (Cambridge, MA: Harvard University Press, 2007).

Dawson, Ashley and Malini Johar Schueller (eds). *Exceptional State: Contemporary US Culture and the New Imperialism* (Durham, NC: Duke University Press, 2007).

Dean, Mitchell. "Foucault, Government, and the Enfolding of Authority," in A. Barry, T. Osborne, and N. Rose (eds), *Foucault and Political Reason: Liberalism, Neo-liberalism, and Rationalities of Government* (Chicago: University of Chicago Press, 1996).

——. "Nomos: Word and Myth," in Louiza Odysseos and Fabio Petito (eds), *The International Political Thought of Carl Schmitt* (London: Routledge, 2007).

Debrix, François. *Re-Envisioning Peacekeeping: The United Nations and the Mobilization of Ideology* (Minneapolis: University of Minnesota Press, 1999).

——. *Tabloid Terror: War, Culture, and Geopolitics* (New York: Routledge, 2008).

Debrix, François and Mark J. Lacy (eds). *The Geopolitics of American Insecurity: Terror, Power, and Foreign Policy* (London: Routledge, 2009).

DeCaroli, Steven. "Boundary Stones: Giorgio Agamben and the Field of Sovereignty," in Matthew Calarco and Steven DeCaroli (eds). *Giorgio Agamben: Sovereignty and Life* (Stanford, CA: Stanford University Press, 2007).

DeLanda, Manuel. *Intensive Science and Virtual Philosophy* (London: Continuum, 2002).

Deleuze, Gilles. *Bergsonism* (New York: Zone Books, 1991).

——. *Difference and Repetition* (New York: Columbia University Press, 1994).

——. "What Is a Dispositif?" in Gilles Deleuze, *Two Regimes of Madness: Texts and Interviews, 1975–95* (New York: Semiotext(e), 2006).

Deleuze, Gilles and Felix Guattari. *Anti-Oedipus: Capitalism and Schizophrenia* (Minneapolis: University of Minnesota Press, 1983).

——. *A Thousand Plateaus: Capitalism and Schizophrenia* (Minneapolis: University of Minnesota Press, 1994).

——. *What Is Philosophy?* (New York: Columbia University Press, 1994).

Derrida, Jacques. "Deconstruction in America: An Interview with Jacques Derrida," *Critical Exchange* Vol. 17 (1985): 1–32.

——. *Given Time: I. Counterfeit Money* (Chicago: University of Chicago Press, 1992).

——. *Politics of Friendship* (New York: Verso, 1997).

——. *Rogues: Two Essays on Reason* (Stanford, CA: Stanford University Press, 2005).

Deuber-Mankowsky, Astrid. "Nothing Is Political, Everything Can Be Politicized: On the Concept of the Political in Michel Foucault and Carl Schmitt," *Telos* Vol. 142 (2008): 135–61.

Diamond, Cora. "The Difficulty of Reality and the Difficulty of Philosophy," in Stanley Cavell, Cora Diamond, John McDowell, Ian Hacking, and Cary Wolfe (eds), *Philosophy and Animal Life* (New York: Columbia University Press, 2009).

Diken, Bulent and Carsten Laustsen (eds). *The Culture of Exception: Sociology Facing the Camp* (London: Routledge, 2005).

Diken, Bulent and Carsten Bagge Laustsen. "The Camp," *Geografiska Annaler* Vol. 88B, No. 4 (2006): 443–52.

Dillon, Michael. "Correlating Sovereignty and Biopower," in Jenny Edkins, Veronique Pin-Fat, and Michael Shapiro (eds), *Sovereign Lives: Power in Global Politics* (New York: Routledge, 2004).

——. "Governing through Contingency: The Security of Biopolitical Governance," *Political Geography* Vol. 26, No. 1 (2007): 41–47.

——. "Governing Terror: The State of Emergency of Biopolitical Emergence," *International Political Sociology* Vol. 1, No. 1 (2007): 7–28.

——. "Security, Race and War," in Michael Dillon and Andrew Neal (eds), *Foucault on Politics, Security and War* (New York: Palgrave, 2008).

Dillon, Michael and Andrew Neal (eds). *Foucault on Politics, Security and War* (New York: Palgrave, 2008).

Dillon, Michael and Julian Reid. *The Liberal Way of War: Killing to Make Life Live* (London: Routledge, 2009).

Dillon, Michael and Luis Lobo-Guerrero. "Biopolitics of Security in the 21st Century: An Introduction," *Review of International Studies* Vol. 34, No. 2 (2008): 265–92.

Disch, Lisa. *Hannah Arendt and the Limits of Philosophy* (Ithaca, NY: Cornell University Press, 1994).

Doty, Roxanne Lynn. "States of Exception on the Mexico-US Border: Security, 'Decisions,' and Civilian Border Patrols," *International Political Sociology* Vol. 1, No. 2 (2007): 113–37.

"Drone Crew Blamed in Afghan Civilian Deaths," *USA Today*, May 29, 2010, available at http://www.usatoday.com/news/world/afghanistan/2010-05-29-afghanistan-civilian-deaths_N.htm (accessed on April 21, 2011).

Duffield, Mark. *Development, Security, and Unending War: Governing the World Peoples* (Cambridge: Polity, 2007).

Dyer-Witheford, Nick and Greig de Peuter, *Games of Empire: Global Capitalism and Video Games* (Minneapolis: University of Minnesota Press, 2009).

Dyzenhaus, David. "Putting the State Back in Credit," in Chantal Mouffe (eds), *The Challenge of Carl Schmitt* (New York: Verso, 1999).

Edkins, Jenny. "Whatever Politics," in Matthew Calarco and Steven DeCaroli (eds), *Giorgio Agamben: Sovereignty and Life* (Stanford, CA: Stanford University Press, 2007): 70–91.

Edkins, Jenny, Veronique Pin-Fat, and Michael J. Shapiro (eds). *Sovereign Lives: Power in Global Politics* (New York: Routledge, 2004).

Elden, Stuart. *Terror and Territory: The Spatial Extent of Sovereignty* (Minneapolis: University of Minnesota Press, 2009).

Esposito, Roberto. *Bios: Biopolitics and Philosophy*, trans. Timothy Campbell (Minneapolis: University of Minnesota Press, 2008).

Fassin, Didier. "Another Politics of Life is Possible," *Theory, Culture & Society* Vol. 26, No. 5 (2009): 44–60.

——. "Ethics of Survival: A Democratic Approach to the Politics of Life," *Humanity: An International Journal of Human Rights, Humanitarianism, and Development* Vol. 1, No. 1 (2010): 81–95.

Foucault, Michel. *Madness and Civilization: A History of Insanity in the Age of Reason* (New York: Vintage Books, 1965).

——. *The History of Sexuality, Vol. 1* (New York: Vintage Books, 1978).

——. *Discipline and Punish: The Birth of the Prison* (New York: Vintage Books, 1979).

——. "Governmentality," in G. Burchell, C. Gordon, and P. Miller (eds), *The Foucault Effect: Studies in Governmentality* (Chicago: University of Chicago Press, 1991).

——. *"Society Must Be Defended": Lectures at the College de France, 1975–76* (New York: Picador, 2003).

——. *Security, Territory, Population: Lectures at the Collège de France, 1977–78* (New York: Palgrave, 2007).

Galli, Carlo. "On War and the Enemy," *CR: The New Centennial Review* Vol. 9, No. 2 (2009): 195–220.

——. *Political Spaces and Global War* (Minneapolis: University of Minnesota Press, 2010).

Garzon, Juan Carlos. *Mafia & Co: The Criminal Networks in Mexico, Brazil, and Colombia* (Washington: Woodrow Wilson International Center for Scholars, first published in Spanish in 2008; no year given for English translation), available at www.wilsoncenter.org (accessed on April 21, 2011).

George, Larry N. "American Insecurities and the Ontopolitics of US Pharmacotic Wars," in François Debrix and Mark J. Lacy (eds), *The Geopolitics of American Insecurity: Terror, Power, and Foreign Policy* (London: Routledge, 2009).

Girard, René. *Violence and the Sacred* (Baltimore, MD: Johns Hopkins University Press, 1977).

——."Sacrifice as Sacral Violence and Substitution," in James G. Williams (ed.), *The Girard Reader* (New York: Crossroad, 1996).

Gourevitch, Philip. *We Wish to Inform You That Tomorrow We Will Be Killed with Our Families: Stories from Rwanda* (New York: Picador, 1999).

Gregory, Derek. *The Colonial Present: Afghanistan, Palestine, Iraq* (Oxford: Blackwell, 2004).

——. "The Black Flag: Guantanamo Bay and the Space of Exception," *Geografiska Annaler* Vol. 88B, No. 4 (2006): 405–27.

——. "Vanishing Points: Law, Violence, and Exception in the Global War Prison," in Derek Gregory and Allan Pred (eds), *Violent Geographies* (New York: Routledge, 2007).

——. "The Biopolitics of Baghdad: Counterinsurgency and the Counter-City," *Human Geography* Vol. 1 (2008): 6–27.

Gregory, Derek and Allan Pred (eds). *Violent Geographies: Fear, Terror, and Political Violence* (New York: Routledge, 2007).

Guillermoprieto, Alma. "The Murders of Mexico," *The New York Review of Books*, October 28, 2010, available at www.nybooks.com/archives/2010/oct/28/murderers-mexico/ (accessed on December 6, 2010).

Habermas, Jürgen. "Hannah Arendt's Communications Concept of Power," in Steven Lukes (ed.), *Power: Readings in Social and Political Theory* (New York: New York University Press, 1986).

Hardt, *Michael and Antonio Negri. Empire* (Cambridge, MA: Harvard University Press, 2001).

——. *Multitude: War and Democracy in the Age of Empire* (New York: Penguin, 2004).

Harvey, David. *The New Imperialism* (Oxford: Oxford University Press, 2005).

Herf, Jeffrey. *Reactionary Modernism: Technology, Culture, and Politics in Weimar and the Third Reich* (Cambridge: Cambridge University Press, 1984).

Hirst, Paul. "Carl Schmitt's Decisionism," in Mouffe (ed.), *The Challenge of Carl Schmitt* (New York: Verso, 1999).

Hobbes, Thomas. *Leviathan* (New York: Penguin, 1985).

——. *On the Citizen* (Cambridge: Cambridge University Press, 1998).

——. *The Elements of Law, Natural and Politic: Part I, Human Nature; Part II, De Corpore Politico; with Three Lives* (Oxford: Oxford University Press, 1999).

Honig, Bonnie. *Political Theory and the Displacement of Politics* (Ithaca, NY: Cornell University Press, 1993).

——. *Emergency Politics: Paradox, Law, Democracy* (Princeton, NJ: Princeton University Press, 2009).

Huysmans, Jef. "International Politics of Exception: Competing Visions of International Political Order between Law and Politics," *Alternatives* Vol. 31 (2006): 135–65.

——. *The Politics of Insecurity: Migration and Asylum in the EU* (London: Routledge, 2006).

Joxe, Alain. *Empire of Disorder* (Los Angeles: Semiotext(e), 2002).

Kalyvas, Andreas. "From the Act to the Decision: Hannah Arendt and the Question of Decisionism," *Political Theory* Vol. 32 (2004): 320–46.

Kearns, Gerry. "Bare Life, Political Violence, and the Territorial Structure of Britain and Ireland," in Derek Gregory and Alan Pred (eds), *Violent Geographies: Fear, Terror, and Political Violence* (New York: Routledge, 2007).

Kervégan, Jean-François. "Carl Schmitt and 'World Unity'," in Chantal Mouffe (eds), *The Challenge of Carl Schmitt* (New York: Verso, 1999).

Koskenniemi, Martti. "International Law as Political Theology: How to Read Nomos der Erde," *Constellations* Vol. 11, No. 4 (2004): 492–511.

Kristeva, Julia. *Powers of Horror: An Essay on Abjection* (New York: Columbia University Press, 1982).

Legg, Stephen (ed.). *Spatiality, Sovereignty, and Carl Schmitt: Geographies of the Nomos* (London: Routledge, 2011).

Levi, Primo. *The Drowned and the Saved* (New York: Vintage International, 1989).

——. *Si c'est un Homme* (Paris: Pocket Juliard, 2008).

Löwith, Karl. "The Occasional Decisionism of Carl Schmitt," in Richard Wolin (ed.), *Martin Heidegger and European Nihilism* (New York: Columbia University Press, 1995).

Luke, Timothy W. "The Discipline of Security Studies and the Codes of Containment: Learning from Kuwait," *Alternatives* Vol. 16, No. 3 (1991): 139–54.

Mansfield, Nick. *Theorizing War: From Hobbes to Badiou* (New York: Palgrave Macmillan, 2008).

Martin, Lauren. "Bombs, Bodies, and Biopolitics: Securitizing the Subject at the Airport Security," *Social and Cultural Geography* Vol. 11, No. 1 (2010): 17–34.

Massumi, Brian. "Everywhere You Want to Be: Introduction to Fear," in Brian Massumi (ed.), *The Politics of Everyday Fear* (Minneapolis University of Minnesota Press, 1993).

Mbembe, Achille. "Necropolitics," *Public Culture* Vol. 15, No. 1 (2003): 11–40.

——. "On Politics as a Form of Expenditure," in John L. Comaroff and Jean Comaroff (eds), *Law and Disorder in the Postcolony* (Chicago: University of Chicago Press, 2006).

McClintock, Anne. "Paranoid Empire: Specters from Guantánamo and Abu Ghraib," *Small Axe* Vol. 13, No. 1 (2009): 50–74.

Meier, Heinrich. *The Lesson of Carl Schmitt: Four Chapters on the Distinction between Political Theology and Political Philosophy* (Chicago: University of Chicago Press, 1998).

"Mexico's Drug War: Falling Kingpins, Rising Violence", *The Economist*, December 16, 2010, available at www.economist.com/node/17733279?story_id=17733279&CFID=157801906&CFTOKEN=30578643 (accessed on April 21, 2011).

Mills, Catherine. "Biopolitics, Liberal Eugenics, and Nihilism," in Matthew Calarco and Steven DeCaroli (eds), *Giorgio Agamben: Sovereignty and Life* (Stanford, CA: Stanford University Press, 2007).

Minca, Claudio. "The Return of the Camp," *Progress in Human Geography* Vol. 29, No. 4 (2005): 405–12.

——. "Giorgio Agamben and the New Biopolitical Nomos," *Geografiska Annaler* Vol. 88, No. 4 (2006): 387–403.

Morton, Stephen and Stephen Bygrave (eds). *Foucault in an Age of Terror: Essays on Biopolitics and the Defence of Society* (London: Palgrave, 2008).

Mouffe, Chantal. "Carl Schmitt and the Paradox of Liberal Democracy," in Chantal Mouffe (ed.), *The Challenge of Carl Schmitt* (New York: Verso 1999).

Nancy, Jean-Luc. *The Experience of Freedom* (Stanford, CA: Stanford University Press, 1993).

Neal, Andrew. "Cutting Off the King's Head: Foucault's Society "*Society Must Be Defended*" and the Problem of Sovereignty," *Alternatives* Vol. 29 (2004): 373–98.

Nealon, Jeffrey T. *Foucault Beyond Foucault: Power and Its Intensifications since 1984* (Stanford, CA: Stanford University Press, 2008).

Nelson, Scott. *Sovereignty and the Limits of the Liberal Imagination* (London: Routledge, 2009).

Neocleous, Mark. *Critique of Security* (Montreal: McGill-Queen's University Press, 2008).

Ojakangas, Mika. *A Philosophy of Concrete Life: Carl Schmitt and the Political Thought of Late Modernity* (New York: Peter Lang, 2006).

——. "A Terrifying World without an Exterior," in Louiza Odysseos and Fabio Petito (eds), *The International Political Thought of Carl Schmitt* (London: Routledge, 2007).

——. "Carl Schmitt and the Sacred Origins of Law," *Telos* No. 157 (Summer 2009): 34–54.

Odysseos, Louiza. "Crossing the Line? Carl Schmitt on the 'Spaceless Universalism' of Cosmopolitanism and the War on Terror," in Louiza Odysseos and Fabio Petito (eds), *The International Political Thought of Carl Schmitt* (London: Routledge, 2007).

O'Kane, David and Tricia Redeker Hepner (eds). *Biopolitics, Militarism, and Development: Eritrea in the Twenty-First Century* (Oxford: Berghan Books, 2009).

Ong, Aihwa. *Neoliberalism as Exception: Mutations in Citizenship and Sovereignty* (Durham, NC: Duke University Press, 2006).

Palladino, Paolo. "Revisiting Franco's Death: Life and Death and Biopolitical Governmentality," in Michael Dillon and Andrew Neal (eds), *Foucault on Politics, Security and War* (London: Palgrave, 2008).

Pan, David. "Carl Schmitt on Culture and Violence in the Political Decision," *Telos* Vol. 142 (2008): 49–72.

Patton, Paul. "Agamben and Foucault on Biopower and Biopolitics," in Matthew Calarco and Steven DeCaroli (eds), *Giorgio Agamben: Sovereignty and Life* (Stanford, CA: Stanford University Press, 2007).

Paye, Jean-Claude. *Global War on Liberty* (New York: Telos Press, 2007).

Petito, Fabio. "Against World Unity: Carl Schmitt and the Western-centric and Liberal Global Order," in Louiza Odysseos and Fabio Petito (eds), *The International Political Thought of Carl Schmitt (*London: Routledge, 2007).

Phillips, A. A. and M. M. Willcock (eds). *Xenophon and Arrian on Hunting* (New York: Aris and Phillips, 1999).

Quéau, Philippe. *Le Virtuel: Vertus et Vertiges* (Paris: Champ Vallon, 1994).

Rasch, William. "Conflict as a Vocation: Carl Schmitt and the Possibility of Politics," *Theory, Culture, and Society* Vol. 17 (2000): 1–32.

Reid, Julian. *The Biopolitics of the War on Terror: Life Struggles, Liberal Modernity, and the Defense of Logistical Societies* (Manchester: Manchester University Press, 2006).

——. "The Biopolitics of American Security Policy in the Twenty-First Century," in François Debrix and Mark J. Lacy (eds), *The Geopolitics of American Insecurity: Terror, Power and Foreign Policy* (London: Routledge, 2009).

Resnais, Alain. *Nuit et Brouillard* (1955) [film].

Robin, Corey. *Fear: The History of a Political Idea* (Oxford: Oxford University Press, 2004).

Rose, Nikolas. "Governing 'Advanced' Liberal Democracies," in A. Barry, T. Osborne, and N. Rose (eds), *Foucault and Political Reason: Liberalism, Neo-liberalism, and Rationalities of Government* (Chicago: University of Chicago Press, 1996).

——. *The Politics of Life Itself: Biomedicine, Power, and Subjectivity in the Twenty-First Century* (Princeton, NJ: Princeton University Press, 2007).

Schaap, Andrew. "Political Theory and the Agony of Politics," *Political Studies Review* Vol. 5 (2007): 56–74.

Schlosser, Kolson. "Bio-Political Geographies," *Geography Compass* Vol. 2, No. 5 (2008): 1621–34.

Schmitt, Carl. *Glossarium—Aufziechnunger des Jahre 1947–51* (Berlin: Duncker & Humblot, 1988).

——. *Über die Drei Arten des Rechtswissenschaftlichen Denkens* (Berlin: Duncker & Humblot, 1993).

——. *The Concept of the Political* (Chicago: University of Chicago Press, 1996).

——. "Ethic of State and Pluralistic State," in Chantal Mouffe (eds), *The Challenge of Carl Schmitt* (New York: Verso, 1999).

——. *The Nomos of the Earth in the International Law of the Jus Publicum Europaeum* (New York: Telos Press, 2003).

——. *Political Theology: Four Chapters on the Concept of Sovereignty* (Chicago: University of Chicago Press, 2005).

——. "Nomos—Nahme—Name," in C. Schmitt, *The Nomos of the Earth in the International Law of the Jus Publicum Europaeum* (New York: Telos, 2006).

——. *Theory of the Partisan: Intermediate Commentary on the Concept of the Political* (New York: Telos Press, 2007).

Secor, Anna. "An Unrecognizable Condition Has Arrived: Law, Violence, and the State of Exception in Turkey," in Derek Gregory and Allan Pred (eds), *Violent Geographies* (New York: Routledge, 2007).

Shapiro, Kam. *Sovereign Nations, Carnal States* (Ithaca, NY: Cornell University Press, 2003).

Shapiro, Michael J. *Methods and Nations: Cultural Governance and the Indigenous Subject* (New York: Routledge, 2004).

——. *Cinematic Geopolitics* (London: Routledge, 2009).

Shukin, Nicole. *Animal Capital: Rendering Life in Biopolitical Times* (Minneapolis: University of Minnesota Press, 2009).

Singer, P. W. *Wired for War: The Robotics Revolution and Conflict in the 21st Century* (New York: Penguin, 2009).

Smith, Neil. *The Endgame of Globalization* (New York: Routledge, 2004).

Snyder, Timothy. *Bloodlands: Europe Between Hitler and Stalin* (New York: Basic Books, 2010).

Sontag, Susan. *Regarding the Pain of Others* (New York: Farrar, Strauss and Giroux, 2003).

Strauss, Leo. *The Political Philosophy of Hobbes: Its Basis and Its Genesis* (Chicago: University of Chicago Press, 1984).

——. "Notes on Carl Schmitt, The Concept of the Political," in Carl Schmitt, *The Concept of the Political* (Chicago: University of Chicago Press, 1996).

Strong, Tracy. "Foreword: The Sovereign and the Exception," in *Carl Schmitt, Political Theology: Four Chapters on the Concept of Sovereignty* (Chicago: University of Chicago Press, 2005).

Taussig, Michael. *Mimesis and Alterity: A Particular History of the Senses* (New York: Routledge, 1993).

van Agtmael, Peter. *2nd Tour, Hope I don't Die* (Portland, OR: Photolucida, 2009).

Vaughan-Williams, Nick. "The Shooting of Jean Charles de Menezes: New Border Politics?" *Alternatives* Vol. 32, No. 2 (2007): 177–95.

——. "Borders, Territory, Law," *International Political Sociology* Vol. 2, No. 4 (2008): 322–38.

——. *Border Politics: The Limits of Sovereign Power* (Edinburgh: Edinburgh University Press, 2009).

——. "Virtual Border (In)security," paper presented at the 2009 annual meeting of the International Studies Association, New York, February 15–18, 2009.

Villa, Dana. *Arendt and Heidegger: The Fate of the Political* (Princeton, NJ: Princeton University Press, 1996).

Vulliamy, Ed. "The Terror," *Vanityfair*, October 21, 2010, available at http://www.vanityfair.com/politics/features/2010/10/drug-wars-in-mexico-201010?printable=true (accessed on April 21, 2011).

——. *Amexica: War Along the Borderline* (New York: Farrar, Straus and Giroux, 2010).

Waldrop, Murray. "Unmanned Drones Could Be Banned, Says Senior Judge," *The Daily Telegraph,* July 6, 2009, available at /www.telegraph.co.uk/news/newstopics/politics/defence/5755446/Unmanned-drones-could-be-banned-says-senior-judge.html (accessed on April 21, 2011).

Walker, R. B. J. *Inside/Outside: International Relations as Political Theory* (Cambridge: Cambridge University Press, 1993).

Weber, Samuel. *Targets of Opportunity: On the Militarization of Thinking* (New York: Fordham University Press, 2005).

Webster's Dictionary (New York: Merriam-Webster, 1953).

Weiler, Gershon. *From Absolutism to Totalitarianism: Carl Schmitt on Thomas Hobbes* (London: Hollowbrook, 1994).

Weisband, Edward. "On the Aporetic Borderlines of Forgiveness: Bereavement as a Political Form," *Alternatives* Vol. 34, No. 4 (2009): 359–81.

Whitehall, Geoffrey. "The Aesthetic Emergency of the Avian Flu Affect," in François Debrix and Mark Lacy (eds), *The Geopolitics of American Insecurity* (London: Routledge, 2009).

Williams, Michael C. *The Realist Tradition and the Limits of International Relations* (Cambridge: Cambridge University Press, 2005).

Wilson, Andrew. *Virtual Politics: Faking Democracy in the Post-Soviet World* (New Haven, CT: Yale University Press, 2005).

Wolfe, Cary. *What Is Posthumanism?* (Minneapolis: University of Minnesota Press, 2010).

Wolin, Richard. "Carl Schmitt: The Conservative Revolutionary Habitus and the Aesthetics of Horror," *Political Theory* Vol. 20 (1992): 424–47.

Wolin, Shelson. *Democracy Incorporated: Managed Democracy and the Specter of Inverted Totalitarianism* (Princeton, NJ: Princeton University Press, 2008).

Youde, Jeremy. *Biopolitical Surveillance and Public Health in International Politics* (London: Palgrave, 2010).

Žižek, Slavoj. "Carl Schmitt and the Age of Post-Politics," in Mouffe (eds), *The Challenge of Carl Schmitt* (New York: Verso, 1999).

——. *In Defense of Lost Causes* (New York: Verso, 2009).

Zolo, Danilo. *Victor's Justice: From Nuremberg to Baghdad* (London: Verso, 2009).

Index